Library of
Davidson College

Results and Problems in Cell Differentiation

A Series of Topical Volumes in Developmental Biology

5

Advisory Board

R. Brown, Edinburgh · F. Duspiva, Heidelberg · H. Eagle, Bronx · M. Feldman, Rehovoth
P. R. Gross, Cambridge · H. Harris, Oxford · H. Holtzer, Philadelphia · W. A. Jensen,
Berkeley · R. J. Neff, Nashville · D. H. Northcote, Cambridge · C. Pelling, Tübingen
W. Sachsenmaier, Innsbruck · H. A. Schneiderman, Irvine · E. Schnepf, Heidelberg
H. G. Schweiger, Wilhelmshaven · G. L. Stebbins, Davis · C. Stern, Berkeley · H. Tiedemann,
Berlin · W. Trager, New York · R. Weber, Bern · R. Wollgiehn, Halle · T. Yamada, Oak Ridge

Editors

W. Beermann, Tübingen · J. Reinert, Berlin · H. Ursprung, Zürich

The Biology of Imaginal Disks

Edited by H. Ursprung and R. Nöthiger, Zürich

With contributions of

J. W. Fristrom, Berkeley · A. García-Bellido, Madrid
W. Gehring, New Haven · R. Nöthiger, Zürich
H. Oberlander, Gainesville · H. Ursprung, Zürich

With 56 Figures

Springer-Verlag New York · Heidelberg · Berlin 1972

ISBN 0-387-05785-4 Springer-Verlag New York Heidelberg Berlin
ISBN 3-540-05785-4 Springer-Verlag Berlin Heidelberg New York

This work is subject to copyright. All rights are reserved, whether the whole or part of the material is concerned, specifically those of translation, reprinting, re-use of illustrations, broadcasting, reproduction by photocopying machine or similar means, and storage in data banks.
Under § 54 of the German Copyright Law, where copies are made for other than private use, a fee is payable to the publisher, the amount of the fee to be determined by agreement with the publisher.

The use of general descriptive names, trade names, trade marks etc. in this publication, even if the former are not especially identified, is not to be taken as a sign that such names, as understood by the Trade Marks and Merchandise Marks Act, may accordingly be used freely by anyone.

© by Springer-Verlag Berlin · Heidelberg 1972. Library of Congress Catalog Card Number 72-75723.
Printed in Germany. Typesetting, printing and bookbinding: Brühlsche Universitätsdruckerei Gießen.

(Photograph by H. Ursprung)

For the 70th Birthday of Ernst Hadorn

> And still they gazed and still the wonder grew
> That one small head could carry all he knew
>
> OLIVER GOLDSMITH

It is indeed an honor to pay tribute to ERNST HADORN on his seventieth birthday. We are pleased and gratified for the opportunity to express our admiration for this unique man and for his many fundamental contributions to modern science.

The year was 1937. That is three and a half decades ago, the publication date of HADORN's „Habilitationsschrift", which dealt with the relation between nucleus and cytoplasm in the development of amphibian hybrid merogones. In these early years of his career he worked exclusively with amphibians. The „Habilitations" contribution was actually a continuation of his hybrid merogone studies. The latter were begun around 1930 and established HADORN as one of the most promising young embryologists. These early papers already carry certain unmistakable trademarks which later characterize and confer such distinction to his work: namely, his uncanny ability to use new methods in significantly different and excitingly new combinations and thus to ask specific new questions from the biological material. Moreover, the main thrust of the merogone experiments had, of course, been directed towards solving the important question of how the genetic material controls development. This theme has never left HADORN and it became, in one way or another, the leitmotiv for all his subsequent scientific work.

In the late 1930's HADORN spent some time in America and began to work with Drosophila. He must have recognized already then the extraordinary potential of this animal for the study of developmental genetics. First, he learned the method of organ transplantation, a technique just introduced for Drosophila. Within a short time, he demonstrated that puparium formation in this fly was brought about by a hormone and that this hormone was released by an endocrine gland – the ring gland. With this discovery HADORN had made a major contribution to the then new and exciting field of insect endocrinology, and he was deeply and permanently involved with Drosophila.

HADORN and his school have produced over the last thirty years an amazing amount of information dealing in general with questions concerning the development and the developmental genetics of Drosophila. In the foreground of all this work stand his contributions to the problem of organ determination. Organ disks were used for these studies. The choice of the test object was perfect. Not only are the external organs of insects composed of their organ-specific component parts, but also each part can be divided into subunit parts that bear in definite order a variety of chitinous ridges, bristles, hairs, and other structures that can be easily recognized as belonging to a definite organ region. Thus, the entire organ can be visualized as a mosaic of areas that are foreshadowed in the presumptive organ anlage from which the adult structures arise. Because the disks have the ability to withstand severe experimental manipulation, including defect operations, and because the adult structures exhibit so much detailed differentiation, the disks present a unique material for the study of the problem of determination. Yet all this would not have been enough. It was HADORN and his students who, by careful observation and with the help of genetic markers, were able to identify accurately the different organ areas by their chitinous structures, and thus made the Drosophila disk the important tool it is today. HADORN's investigations on the regulation and determination of the Drosophila genital disk, for instance, have become a classic and are required reading for all students of embryology. There are, of course, many other examples with like implications.

It has been said, and rightly so, that in order to advance a biological field, or even a specific area, one must bring to bear on it as many different techniques as possible. This is one of the reasons why HADORN's papers are so exciting. He has attacked his problems by genetical, biochemical, radiological, nutritional, tissue culture, organ culture, transplantation, and other techniques. Such methods have yielded much new fundamental knowledge, especially in the fields of development and developmental genetics, as these are the two areas of his primary interest. By a constant production of original research of superb quality and vision HADORN has become one of the few great embryologists of our time. And yet HADORN's greatest contribution to embryology has not even been mentioned. It occurred when, in the early sixties, he discovered — again working with imaginal disks of Drosophila — the phenomenon he termed transdetermination. It means that an imaginal disk, for instance a genital disk, which is expected to differentiate according to its original state of determination only into genital structures, also can develop structures characteristic of other organs or body regions, such as wing, leg, or thoracic parts. This change from a given state of determination to a different one is called transdetermination. These findings are even more remarkable if one considers that they occurred in Diptera — a material which embryologists considered to be the prototype for complete determinative development. Its is true that from time to time we had some glimpses that early, and supposedly final, determinative events did not always carry their rigid state through later developmental periods, as indicated by the regulative capacity of rather old organ disks. We have also long known of the existence of the phenomenon of homöosis, i.e. the replacement of an organ by one belonging to another region of the body. But these "replacement organs" were always of a neighboring organ type and we interpreted their presence as a result of early interactions between neighboring fields. Yet these oddities never really made us question our concept of determination. HADORN's experiments did! Their impact was immediate for they did what we had not done; they forced us to rethink our notion of determination. The embryological perspectives inherent in the general problem of determination widened and radiated in many directions putting new and fundamental issues into focus, especially those relating to genetic aspects. HADORN and his students could now see the possibility of seeking a deeper understanding of the determination processes by approaching the problem on the molecular level.

Thus, HADORN has led the Drosophila disks through classical to biochemical and finally to molecular genetics — and this is a long way. As grateful and as proud as these disks must have been for the opportunity to contribute so much to science on their long journey, so are we to the man who allowed us to participate in these travels.

One must mention yet another great gift from HADORN, one which illustrates perhaps better than anything else the breadth of his knowledge and his analytical ability. It is his book *Letalfaktoren* which appeared in 1955 in German and was translated into English in 1961 as *Developmental Genetics and Lethal Factors*. A comprehensive overview of the field, it represents the best synthesis of classical developmental genetics available anywhere.

One last word: perhaps the most distinct characteristics of HADORN's personality are his absolute devotion to and infectious enthusiasm for his work, as well as his relentless energy. He is an inspiring teacher and a superb lecturer, an understanding and stimulating colleague, and a good man.

From all your friends, students, and colleagues from many lands — *Happy Birthday*, ERNST HADORN! We all wish that for many years to come you will continue to provide us with leadership, inspiration, and council in our search for new vistas in embryology and genetics.

DIETRICH BODENSTEIN

Preface

No working hypothesis amounts to much until it has been tested on suitable material. Indeed, the choice of an appropriate experimental system has often been the key to the solution of a problem. The present volume is devoted to insect imaginal disks. These groups of larval cells are the primordia of precisely characterized adult counterparts, without apparent function in larvae. At the onset of metamorphosis, the subtle interplay of hormonal signals brings growth to a halt, and differentiation begins. In the fruitfly, a host of mutations are known to affect the development of disks; these provide ample material for analysis.

It was largely ERNST HADORN's ingenuity that directed the attention of many scientists around the world to this promising experimental system, and to him this volume is dedicated. All the contributors have been associated with him at one time or another, as graduate students, postdoctoral fellows, or colleagues.

Each author has attempted to cover comprehensively the topic assigned to him. This has inevitably led to some overlapping, for which the editors should be blamed, not the authors, as this results from the way the topic was subdivided at the outset.

We believe this volume will be a welcome sourcebook for the specialist in the field, and a provocative monograph for the uninitiated scientist interested in the exciting area of cell determination.

Tübingen, Berlin, Zürich
April 1972

W. BEERMANN, R. NÖTHIGER,
J. REINERT, H. URSPRUNG

Contents

The Larval Development of Imaginal Disks
by Rolf Nöthiger

I. Introduction.	1
II. Determination: Operational Definition and Criteria	3
III. Techniques of Analysis	4
A. Transplantation.	4
B. Dissociation and Reaggregation.	4
C. Genetic Mosaics.	5
IV. Singling-Out and Programming the Cells of the Imaginal Disks	8
A. Time of Singling-Out	8
B. Determining Factors.	9
C. Number of Primitive Cells	10
D. Role of Disks in Larval Life	12
V. Growth and Morphogenesis.	12
A. Dynamics of Growth	12
1. Histological Analysis	12
2. Induced Mitotic Recombination	13
3. Regional Differences in Mitotic Activity	14
4. Criticism	15
B. Morphogenesis.	15
VI. The Switch from Cell Multiplication to Cell Differentiation	18
A. Competence	18
B. The Acquisition of Competence.	19
1. Humoral Factors	19
2. Mutations Affecting Competence	20
3. Causal Analysis.	20
VII. The Acquisition and Propagation of a Given State of Determination	21
A. Individual Disks and Parts of Disks are Determined to Develop into Specific Regions of the Adult Fly	21
B. In Late Third Instar Larvae, Even Single Cells are Determined to Develop into the Cell Types of a Specific Region of the Adult Fly	21
C. Determination Proceeds Stepwise	23
D. Determination has Reached a "Prefinal" State at the End of Larval Development.	23

E. Determination Occurs in Groups of Cells 26
　　F. The State of Determination in Disks of Mature Larvae is Very Stable Though Not Irreversible. It is Usually Propagated Conservatively to Daughter Cells 28
　　References 28

The Stability of the Determined State in Cultures of Imaginal Disks in Drosophila
by WALTER GEHRING

　　I. Determination and Differentiation 35
　II. Imaginal Disks and their Culture in vivo 36
　III. Determination of Imaginal Disks and their Progenitor Cells 39
　IV. Mirror-Image Duplication and Regeneration 44
　　V. Cell Heredity and Transdetermination 45
　VI. Clonal Analysis 47
　VII. Possible Causes of Transdetermination 52
VIII. Conclusions 54
　　References 55

Pattern Formation in Imaginal Disks
by A. GARCÍA-BELLIDO

　　I. Introduction 59
　II. Pattern Variations 60
　　A. Genetic Control of Patterns 60
　　B. Pattern Mutants 63
　　C. Experimentally Induced Variations of Patterns 65
　III. The Arising of Topological Differences 66
　IV. The Final Pattern 70
　　A. Cell Lineage 71
　　B. Cell Affinities 76
　　C. Cell Interactions 82
　　V. Concluding Remarks 85
　　References 86

The Fine Structure of Imaginal Disks
by HEINRICH URSPRUNG

　　I. Introduction 93
　II. Fine Structural Organization of Late Third Instar Larval Imaginal Disks 94
　III. Developmental Changes in Disk Fine Structure 98

A. Epithelial Cells 98
  B. Adepithelial Cells 101
IV. Disk Eversion 101
V. Cell to Cell Communication and Adhesion 101
VI. Fine Structure of Mutant Imaginal Disks 103
  A. Cell Death 103
  B. "Neoplasms" 104
VII. Non-autonomous Development of Disks: Ultrastructural Evidence for Uptake of Materials from Larval Hemolymph 105
VIII. Virus-like Particles 105
IX. Conclusions 105
  References 106

The Biochemistry of Imaginal Disk Development
by JAMES W. FRISTROM

I. Introduction 109
II. Methodology 110
  A. The Preparative Isolation of Imaginal Disks 110
  B. Culture Conditions 113
III. Biochemical Composition of Imaginal Disks 115
  A. DNA 115
  B. RNA 116
  C. Protein 116
IV. Physiology and Metabolism of Insect Hormones 116
  A. Ecdysones 116
  B. Juvenile Hormones 120
V. Macromolecular Synthesis in Disks 121
  A. General Comments 121
  B. DNA Synthesis 121
  C. RNA Synthesis 122
    1. Kinetics of Precursor Incorporation 122
    2. Base Composition of Newly Synthesized RNA 122
    3. Stability of Newly Synthesized RNA 123
    4. rRNA Synthesis 123
    5. Hybridization Characteristics of Newly Synthesized Disk RNA . 124
  D. The Effect of Ecdysone on RNA Synthesis 126
    1. Rate of Incorporation of Precursors into RNA 126
    2. rRNA Synthesis 127
    3. Hybridization of Disk RNA 129

E. The Effect of Juvenile Hormone Upon RNA Synthesis 132
F. Protein Synthesis 133
G. Effect of Ecdysone on Protein Synthesis. 136
H. Effect of Juvenile Hormone Upon Protein Synthesis 140
VI. Cuticle Formation and Hexosamine Metabolism 140
VII. The Mechanism of Disk Evagination 142
VIII. Regulatory Mechanisms in Disk Metamorphosis 145
References 150

The Hormonal Control of Development of Imaginal Disks
by HERBERT OBERLANDER

I. Introduction 155
II. Hormonal Control of Molting and Metamorphosis 156
   A. Brain Hormone 156
   B. Molting Hormone 156
   C. Juvenile Hormone 157
III. Imaginal Disks as Targets of Ecdysone and Juvenile Hormone 158
   A. In vivo Experiments 158
   B. In vitro Experiments 159
      1. Morphological Effects 159
      2. Competence of the Disks to Respond to Ecdysone 161
      3. Differences between Effects of α-ecdysone and 20-Hydroxyecdysone 161
      4. α-ecdysone: Trigger or Sustained Stimulus 161
      5. α-Ecdysone Cofactors 162
IV. Hormonal Control of Proliferation in Imaginal Disks 162
   A. In vivo Experiments 162
   B. In vitro Experiments 164
V. Hormonal Control of RNA and Protein Synthesis in Imaginal Disks . . . 165
   A. In vivo Experiments 165
   B. In vitro Experiments 166
VI. Conclusions 167
References 168

Contributors

Fristrom, James, W., Dr., Department of Genetics, University of California, Berkeley, California (USA)

García-Bellido, A., Dr., Instituto Genética y Antropología, Centro Investigaciones Biológicas, C.S.I.C., Madrid (Spain)

Gehring, Walter, Dr., School of Medicine, Department of Anatomy, Yale University, New Haven, Connecticut (USA)

Nöthiger, Rolf, Dr., Zoologisch-vergleichend anatomisches Institut der Universität, Zürich (Switzerland)

Oberlander, Herbert, Dr., United States Department of Agriculture, Agricultural Research Service, Gainesville, Florida (USA)

Ursprung, Heinrich, Dr., Laboratorium für Entwicklungsbiologie, Zoologisches Institut der ETH, Zürich (Switzerland)

The Larval Development of Imaginal Disks

ROLF NÖTHIGER

Zoologisches Institut der Universität Zürich, Zürich

I. Introduction

The life of holometabolous insects, such as *Drosophila*, is characterized by two separate and completely different phases of development. Out of the egg hatches a larva which then becomes transformed during metamorphosis into the adult insect, the imago. The future adult is "hidden" within the larva in the form of the so-called imaginal cells. After fertilization a series of syncytial divisions produces a homogeneous population of nuclei (energids) which later move into the cortical periplasm at the egg's periphery where they form the cellular blastoderm (review by ANDERSON, 1972). This is probably the stage at which the presumptive imaginal cells are segregated from the larval cells (see p. 8). These two distinct populations of cells perform their specific vital functions at entirely different periods of development. During metamorphosis, the larval organization breaks down, and the adult insect is formed anew from the imaginal cells. (There are a few exceptional organ systems, e.g. the Malpighian tubules, that persist through metamorphosis.)

Imaginal cells may be assembled into morphological structures, the imaginal disks (see Fig. 1 in GEHRING, this volume). The disks are surrounded by a non-cellular basement lamina and contain, besides the disk epithelium proper, a "peripodial membrane". This is also an epithelium whose function, however, is not yet known (URSPRUNG and SCHABTACH, 1968; POODRY and SCHNEIDERMAN, 1970). Imaginal cells may also be grouped in nests of cells without a peripodial membrane. This is the case for the presumptive cells of the adult abdominal epidermis, the so-called histoblasts. — And finally, imaginal cells may be integrated into the respective larval organ system, e.g. the salivary gland or the gut. This review, however, will not deal with this last class.

Imaginal disks arise in definite numbers and at definite sites during embryogenesis (Fig. 1). Only minor variations exist between different species of Diptera (DÜBENDORFER, 1971). Histological techniques render the disks visible for the first time during late embryonic or early larval development, with significant differences in the time of appearance between different disks and between disks of different species (AUERBACH, 1936; ANDERSON, 1963). The disks are ectodermal in origin, whereby it is not known, how and when, if at all, mesodermal cells enter the disks. They arise by a process of invagination of ectodermal or embryonic epidermal cells. The surrounding cells of the larval epidermis become polytene and increase in volume while in the small imaginal cells DNA-synthesis is followed by mitoses, and the cells of a disk, therefore, increase in number. This is true for *Drosophila*, *Calliphora* and many other Diptera, whereas in *Chironomus*, the larval epidermal cells divide and multiply (FISCHER and ROSIN, 1969). During the phase of cell multiplication the disks acquire their specific and typical morphology. The adult body is formed by the specific mosaic contribution of each disk. Metamorphosis begins under the influence of hormonal stimuli. Cell divisions in the disks soon cease, and dramatic morphogenetic and differentiative events transform the disks into their corresponding imaginal structures (Fig. 1).

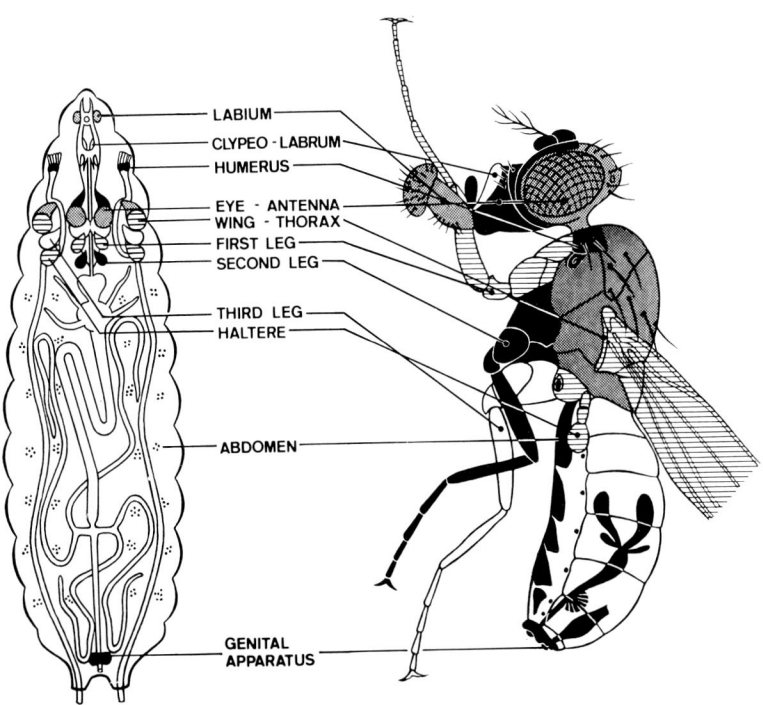

Fig. 1. Schematic representation of the larval organization and the location of the different disks. Disks and their corresponding adult derivatives are connected by lines and given the same hatching or shading (Slightly modified from WILDERMUTH, 1970)

During imaginal differentiation, the cells of the disks are transformed into an astonishing variety of different structures, such as ommatidia with their complex auxiliary apparatus of the compound eye, sensilla which serve as mechanoreceptors or chemoreceptors, bristle organs (bristles or chaetes), and trichomes (hairs, which are simple protrusions of an epidermal cell). Bristles and trichomes are derived from a single cell of a disk (LEES and WADDINGTON, 1942; LAWRENCE, 1966). Their size, shape and arrangement are characteristic for a particular body region (Fig. 1 in GARCIA-BELLIDO, this volume). The cells are beginning to synthesize cuticular proteins, chitin, and eye pigments. This array of different performances is surprising since no differences at the electron microscopic level have been detected so far between the imaginal cells of different disks (URSPRUNG, this volume). An exception are the precursor cells of the ommatidia which become recognizable in white prepupae (WADDINGTON and PERRY, 1960; PERRY, 1968).

The sequence of events described above leads to the formulation of four main groups of problems:

1) The creation of two principally different cell populations, namely larval and imaginal epidermal cells.
   a) When does this separation take place?
   b) What are the controlling factors?
   c) How many primitive cells enter a disk?
   d) Are imaginal disks essential for the larva?

2) The growth and morphogenesis of disks.
   a) What are the dynamics of growth within a disk?
   b) How do disks acquire their typical shape?

3) The switch from cell multiplication to cell differentiation.
   How is this switch brought about, and how is it controlled? What are the conditions required for the switch to take place?

4) The acquisition and propagation of a given state of determination.
   a) When do whole disks, regions, and cells of a disk become determined for a particular developmental pathway?
   b) How does determination proceed?
   c) What is the state of determination in the cells at the end of larval development?
   d) How does determination take place?
   e) How stable is a given state of determination?

Our discussion of these problems will be limited to *Drosophila*, with occasional side glances to other insects.

II. Determination: Operational Definition and Criteria

The term "determination" is purely operational. It is used by embryologists to describe the processes of programming a population of cells for a specific pathway of development by "singling it out from among the various possibilities for which a cellular system is competent" (HADORN, 1965). We imagine that determination acts restrictively, presumably by activating or repressing certain batteries of genes. However, we have as yet no experimental clues to tell us what the molecular processes underlying determination may be. The discrimination between determination and differentiation may even appear artificial, if determination is regarded simply as a step, or a series of steps, in the normal course of development. The imaginal disks, however, exhibit features which require both terms: the determined state of these larval primordia can be experimentally demonstrated (p. 21), but the program of imaginal differentiation (p. 2) is not expressed until metamorphosis. Furthermore, this expression can be delayed indefinitely by culturing disk fragments in adult flies (see p. 4, and GEHRING, this volume). Only when exposed to the hormones of metamorphosis will competent cells (see p. 18) respond with the differentiation of a qualitatively and quantitatively predictable inventory of structures.

The following criteria have been used to ascertain determination in imaginal disks:

a) *Mosaic development:* a disk or a fragment of it, is considered to be determined if, after isolation, it differentiates into those structures that correspond to its prospective fate *in situ;* or conversely, if extirpation of disks, or of cells from a disk, results in specific corresponding defects. Individual cells are determined if they differentiate autonomously when, after dissociation and reaggregation, they happen to be surrounded by cells of a different type.

b) *Properties of specific cell recognition:* the capacity of cells in reaggregates to associate and form integrated mosaic structures when they belong to the same histotype, or to sort out when they belong to different histotypes, is taken to be indicative of cell determination.

c) *Cell heredity:* a given state of determination is usually reproduced faithfully for many cell generations (see GEHRING, this volume).

d) *Cell lineage analysis:* when a clone of genetically marked cells comprises two or more different cell types we infer that their common ancestor cell was not yet determined for any one of the different pathways at the time of clonal initiation.

III. Techniques of Analysis

Due to their size (from 50—500 μ in diameter at the end of the larval period) imaginal disks can easily be manipulated for fragmentation, dissociation and reaggregation, tissue culture and transplantation (see WILT and WESSELLS, 1967, chapters by SCHNEIDER; SCHNEIDERMAN; URSPRUNG).

A. Transplantation

Transplantation of imaginal disks was introduced by EPHRUSSI and BEADLE (1936) to study the differentiation of eye color mutants. The technique proved to be one of the most powerful for the analysis of a variety of developmental problems in *Drosophila*. When competent (see p. 18) disks, fragments of disks, or reaggregated cell clumps are transplanted into larval hosts they will undergo imaginal differentiation synchronously with their host; the implant can then be dissected and the developmental performance examined. Alternatively, they may be subjected to additional cell divisions, literally *ad infinitum* if desired, by transplantation into adult flies whose hormonal milieu allows cell proliferation, but no imaginal differentiation to occur (see GEHRING, this volume). This can be induced any time at will by transplanting the blastema back into a metamorphosing host.

B. Dissociation and Reaggregation

Disks may be dissociated enzymatically (HADORN, ANDERS, URSPRUNG, 1959; GARCIA-BELLIDO, 1966) or mechanically (NÖTHIGER, 1964) into small groups of cells and single cells. With the use of autonomous cell marker mutations, such as *y, w, sn, f, mwh*, etc. (for gene symbols see LINDSLEY and GRELL, 1968) mosaic aggregates of cells from different disks or different regions of a disk can be created. This technique allows us to study problems of cell interaction, pattern formation (see GARCIA-BELLIDO, this volume), and determination (see p. 26).

C. Genetic Mosaics

Other ways of obtaining mosaics are the production of gynandromorphs, the induction of mitotic recombination or somatic mutation, and position-effect variegation. Such genetic mosaics are being used in a variety of organisms including plants (Stewart and Dermen, 1970) and mammals (Tetterborn et al., 1971; Nesbitt, 1971) to study problems of cell interaction and cellular autonomy of mutant expression. The technique permits us to estimate the number of initial progenitor cells of a primordium and to characterize the dynamics of cell multiplication, growth patterns, and morphogenesis (see p. 12, and Garcia-Bellido, this volume).

a) *Gynandromorphs* are mosaically composed of genetically male and female tissue. In *Drosophila*, they arise when one of the two X-chromosomes is lost during development of a female zygote, resulting in XX (female) and XO (male) tissue patches. If the zygote was originally heterozygous for autonomous X-linked cell marker mutations which affect bristle color (e.g. y), bristle shape (e.g. *sn* or *f*), or eye color (e.g. *w*), the mosaicism is recognizable on the body surface of the adult fly because the recessive mutations are uncovered through the loss of the wildtype alleles. Spontaneously, gynandromorphs arise very rarely. But their frequency can be drastically increased if an unstable ring-X chromosome (Hinton, 1955) is used, or if females are homozygous for the mutation *claret-nondisjunctional (ca^{nd})*, or if the paternal X chromosome has been treated with X-rays (Patterson and Stone, 1938). Gynandromorphs as an analytical tool were long restricted to *Drosophila*, but are now also being exploited in *Habrobracon* (Clark, Gould, and Graham, 1971). There is good experimental evidence (discussed by Garcia-Bellido and Merriam, 1969a; Postlethwait and Schneiderman, 1971a; Ripoll, 1972) for the following 4 statements:

1) The loss of one of the X chromosomes occurs in the first, or one of the very first, zygotic divisions; subsequently, the X chromosome becomes stable: gynandromorphs consist, in general, of 50% male and 50% female tissue.

2) The nuclei of the male and female clone do not significantly intermingle: Male and female tissue occupy large, continuous patches.

3) The plane of the first zygotic division (Wald, 1936) or, in any case, the plane separating the male and female clones, is random: the border line between male and female tissue can separate any two parts of the animal. The "textbook-type" of a left-right gynandromorph is just one possibility of how male and female tissue can be separated.

4) Male *(XO)* and female *(XX)* nuclei have equal developmental chances to multiply and to take part in the formation of any given body region: any region or structure on the adult fly is XO or XX equally frequently.

b) *Mitotic recombination*, or *somatic crossing-over*, was first demonstrated convincingly for *Drosophila* by Stern (1936) and has since been used extensively as an elegant and powerful analytical tool (see Stern, 1968). For many years, *Drosophila melanogaster* was the only higher organism, besides certain fungi, for which mitotic recombination was known to occur. Now it is also established for the Soja bean *Glycine max* (Vig and Paddock, 1970) and for the housefly *Musca domestica* (Nöthiger and Düben-

DORFER, 1971). The frequency of mitotic recombination is ordinarily very low. It depends on the genotype and various external parameters. It can be raised to several times the spontaneous level by X-ray treatment of embryos or larvae (BECKER, 1957). The exchange which requires four chromatid strands to be present (Fig. 2) occurs in females as well as in males. Depending on the type of chromosome segregation it may result in the production of two daughter cells each of which will develop into a clone that differs genetically from the common mother cell. The result is a so-called

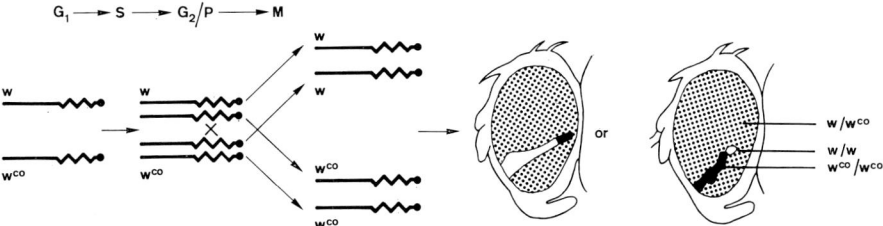

Fig. 2. Mitotic recombination in the X chromosome of *Drosophila*. Proper segregation leads to "twin spots" in the eye. w and w^{co} are different alleles of the *white*-locus. Note the shape of the clones and the difference in size between clones in the anterior and clones in the posterior eye region. G_1, G_2, S, P, M: phases of the cell cycle (After GEHRING and NÖTHIGER, 1972)

"twin spot" (see Fig. 2). It appears that some 70 to 80% of the recombinational events take place in the heterochromatin near the centromere (BECKER, 1969; GARCIA-BELLIDO, 1972). The mechanics of this process are still poorly understood (WALEN, 1964) and are currently intensively being studied (HAENDLE, 1971a, b; MERRIAM and FYFFE, 1972; MERRIAM, NÖTHIGER, and GARCIA-BELLIDO, 1972; MERRIAM and GARCIA-BELLIDO, 1972). X-rays allow us to initiate a genetically marked clone at any time between blastoderm formation and the end of mitotic activity which signals the beginning of imaginal differentiation. (According to BRYANT and SCHNEIDERMAN (1969) no clones could be induced during cleavage. The reason for this failure is not known.) Since mitotic recombination is a cellular event, its frequency will depend on the number of targets, i.e. sensitive cells, whose chromosomes are in a 4-strand stage of the cell cycle at the time of irradiation. The *size* of a developing clone will provide information on the number of cell divisions having passed between initiation of a clone and the last mitoses; and the *shape* of a clone will tell us something about morphogenetic processes.

On strictly formal grounds — but see limitations below —, the number of cells (N_t) present in a disk at the time of irradiation can be estimated if the total number of bristles, ommatidia, or trichomes (N_T) differentiated by this disk is divided by the average number of marked bristles, ommatidia, or trichomes $(\bar{x}_t)$ in a twin spot:

$$N_t = \frac{N_T}{\bar{x}_t}.$$

If the genetic marker configuration allows for single spots only, $\bar{x}_t$ must be multiplied by 2. If the mean spot size is calculated from clones located in all parts of a disk's derivatives, any regional differences in mitotic activity will be averaged out[1].

It becomes immediately clear that this procedure is exact only if each cell of a disk gives rise to a phenotypically recognizable structure (bristle, ommatidium or trichome). It, therefore, will yield a maximum value for the number of cells which corresponds to the number of such structures differentiated by a disk. The method, in fact, can "count" only those cells among whose progeny at least one cell will eventually form a bristle, trichome, or ommatidium. This estimate will approximate the actual cell number when the clones are large, i.e. when they were initiated early in larval life. However, it will become less and less accurate with progressing larval age. This is so because detectable clones will finally comprise only one structure. For the bristles e.g., this stage will the sooner be reached the less dense these are placed. Cell counts made in prepupal disks at the end of the proliferative phase may help to correct for this technical inadequacy.

A further limitation arises from the fact that autonomous cell marker mutations are so far restricted to cuticular structures and eye pigments. Muscles and other soft parts, e.g. the entire inner genitalia, will escape such an analysis.

Contrary to an earlier indication of a possible "delayed effect" (BECKER, 1956), it is now generally believed that the recombinational event takes place at or shortly after irradiation (HAENDLE, 1971a; MERRIAM and FYFFE, 1972). The time of X-ray treatment is, therefore, taken as the time of clone initiation. In analyses with X-ray-induced clones, we assume that the irradiation has no or little side-effects, and that the resulting homozygous cell clones still have a developmental capacity equal to that of their neighboring heterozygous cells. Doses of 1000 R applied to *Drosophila* larvae 72 and 96 h after oviposition did not delay the time of puparium formation, nor increase pupal mortality, nor did they result in morphological aberrations, as compared with non-irradiated controls. However, younger larvae exhibited both a delay in puparium formation and a higher mortality (GARCIA-BELLIDO and MERRIAM, 1971a). SCHWEIZER (1972) also showed for the male genital disk of *Drosophila* that X-ray doses of 2000—4000 R, applied to 68 h old larvae, kill cells and that this loss is subsequently compensated for, or regulated, by an increase in cell divisions. This, in turn, will lead to larger clones than would be formed by normal unirradiated cells. Also SPREIJ (1971) found a significant amount of cell death after 1000 R had been delivered to larvae of *Calliphora* in the middle of the feeding period of the third instar. As in the case just mentioned, this damage is obviously repaired since the morphology of the adult structures is normal. But whereas *Calliphora* larvae respond to a dose of 1000 R with a delay in pupation of 24—48 h, development is apparently normal for *Drosophila* larvae of a comparable age, indicating a higher radiosensitivity of *Calliphora*. A similarly high radiosensitivity was found for larvae of *Musca domestica* (NÖTHIGER and DÜBENDORFER, 1971). It therefore becomes a matter of steering the boat between Skylla and Charybdis so that the chosen X-ray dose will guarantee a

---

[1] A considerably less tedious approximation is possible by planimetrically determining the total area of an imaginal structure (e.g. wing, mesonotum) and dividing it by the average area of the marked clones. Hereby, relative cell densities, if not corrected for, will introduce errors (compare e.g. BRYANT, 1970, and GARCIA-BELLIDO and MERRIAM, 1971a, for clone sizes on the dorsal mesothorax).

reasonable frequency of mitotic exchanges on the one hand, and a minimum cell damage on the other.

c) *Position-effect variegation* can occur in certain chromosomal rearrangements when wildtype euchromatic loci come to lie near a heterochromatic breakpoint. There is a low probability per cell for such a wildtype allele to become non-functional at some time during development. The alternatives "functional" or "non-functional" are stabilized and clonally propagated even in prolonged cultures (HADORN, GSELL, and SCHULTZ, 1970; GSELL, 1971). The phenomenon is not understood (JANNING, 1970), but is nevertheless useful in developmental analyses (review by BAKER, 1968). Restrictions arise from the fact that the time of clonal initiation is out of the experimenter's control.

d) *Final remarks:* There are important differences between mechanically and genetically produced mosaics. Dissociation will disrupt cell contact devices and membrane specialisations, such as septate desmosomes, tight and gap junctions (see URSPRUNG, this volume). These structures as well as any other possible channels of cell communication must reestablish themselves anew upon reaggregation. We do not know how many cells of each genotype form the initial aggregate, and how these cells multiply. In genetically produced mosaics, on the other hand, the cells of different genotypes remain in place and retain their original cell contacts. Upon induced mitotic recombination or somatic mutation, we are reasonably certain that in most cases the mosaic patch arose from a single homozygous cell.

IV. Singling-Out and Programming the Cells of the Imaginal Disks

A. Time of Singling-Out

The time at which some of the embryonic epidermal cells are destined to become imaginal cells and hence to develop into particular adult structures appears to lie around the formation of the cellular blastoderm, i.e. some $2^1/_2$ h after fertilization at 25° C (for a description of the early embryology see COUNCE, 1972; ANDERSON, 1972). This is some 10—20 h before the disks become histologically recognizable and, probably, also precedes the time when larval cells start polytenisation.

Temperature (HENKE and MAAS, 1946) or ether shocks (GLOOR, 1947) applied to early embryos of *Drosophila* can lead to defects or developmental changes which are phenocopies of known homoeotic mutations, such as *tetraptera*. GLOOR (1947) found that only a short period ranging from 1—5 h after oviposition was effective in producing the phenocopies. The maximum sensitivity lay around 3 h with 30% of the treated embryos showing the *tetraptera* phenotype. It is conceivable that the shocking interfered with physiological determinative events occurring at the time of treatment. GEIGY (1931a) was able to induce imaginal defects (duplication or elimination of leg parts) by UV-irradiation of whole embryos: he observed a topographical correlation between the site of irradiation (left-right) and the location of the imaginal defect. However, he got defective adults only if irradiation was performed after $7^1/_2$ h, when the larval organogenesis had already been completed. Earlier stages either did not survive, or developed into normal adults. Similarly, HOWLAND and CHILD (1935) and HOWLAND and SONNENBLICK (1936) who mechanically pricked or injured early embryos by removal of material found that *Drosophila* eggs prior to blastoderm formation exhibited a certain degree of regulation. Localized defects resulted if embryos were treated after blastoderm formation. HADORN et al. (1968) have introduced the technique of mixing parts of embryos from two different genotypes. CHAN and GEHRING (1971) succeeded in refining this technique down to blastoderm stages. Their data indicate that the blastoderm cells are determined at least for anterior and posterior imaginal qualities (see GEHRING, this volume), but the experiments do not yet provide evidence for individual disks to be determined at this early stage.

The method of induced mitotic recombination still permits the most refined analysis of this problem. Genetically marked clones, induced around the time of blastoderm formation, were always confined to the derivatives of one particular disk (BRYANT and SCHNEIDERMAN, 1969; BRYANT, 1970; GARCIA-BELLIDO and MERRIAM, 1971a; POSTLETHWAIT and SCHNEIDERMAN, 1971a). Even very large clones never extended into a neighboring disk. The shortcomings of this technique arise from the fact that we do not know for sure whether all clones were actually initiated at blastoderm formation, and secondly whether the clone did not extend into the larval epidermis which represents "silent regions" in our kind of analysis. However, whereas the latter argument is relevant for the question when imaginal disk cells become separated from larval epidermal cells, the experiments indicate that the different disks are already separated from each other in this early stage.

Nuclear transplantation, on the other hand, revealed that the preblastoderm nuclei are still pluripotent (ILLMENSEE, 1968, 1970; SCHUBIGER and SCHNEIDERMAN, 1971; ZALOKAR, 1971). ILLMENSEE (1972) has recently succeeded in transplanting single nuclei into unfertilized recipient eggs. The donor nuclei were taken from different stages and regions of cleavage and preblastoderm embryos. Developing embryos were cultured in the abdomens of adult females where the prospective imaginal cells proliferated and matured. When transferred into larval hosts, these blastemas could give rise to all cuticular structures of the imago, irrespectively of the region in the embryo from which the donor nuclei were taken.

The data from gynandromorphs (e.g. GARCIA-BELLIDO and MERRIAM, 1969a) lead to the conclusion that the position of the nuclei rather than their genealogy determines their developmental fate. These results still leave open the question whether the preblastoderm nuclei became determined by their surrounding cytoplasm (Hofplasma), or only after immigration into the periplasm. ILLMENSEE's experiments, however, seem to rule out the first of these two alternatives.

B. Determining Factors

Very little is known about such factors. It is conceivable that the periplasm exerts a determining influence on the immigrating cleavage energids (nuclei plus some surrounding cytoplasm). Such an influence is clearly established for the pole cells. The polar plasm, located at the posterior end of the *Drosophila* egg, is rich in ribonucleoprotein complexes (polar granules) (MAHOWALD, 1971). GEIGY (1931b) irradiated the pole plasm with UV and found that surviving flies were sterile, with empty gonads. This finding has recently been confirmed by NÖTHIGER and STRUB (1972) with embryos that were only 2—20 min old. The same determinative effect of the pole plasm was found for *Culex* (OELHAFEN, 1961), and other insects (GEYER-DUSZYNSKA, 1959; BANTOCK, 1970). It is probable that the polar granules are responsible for this effect since they are the most likely UV-targets due to their RNA content. Other attempts to prove the existence of determining factors, or even simply demonstrate a determining role of the periplasm, mostly failed. NÖTHIGER and STRUB (1972) obtained imaginal defects after irradiating small areas of the lateral periplasm in embryos that contained only 2—4 nuclei (2—20 min after oviposition). But no topographical correlation existed between the site of irradiation and the location of the imaginal defect. This does not rule out the occurrence of determining factors in the periplasm. These might appear only later during cleavage; or the technique used may not be adequate.

We shall now present some positive results: KALTHOFF (1971a, b) was able to induce a double abdomen in early cleavage stages of the chironomid *Smittia*. UV-irradiation of the anterior egg regions resulted in 100% of the cases in the formation of a second abdomen symmetrical to the posterior one. This experiment shows that an external stimulus on the periplasm prior to nuclear immigration can reproducibly change the prospective developmental fate of the nuclei. KALTHOFF (1971b) used a main wave length of 260 nm which suggests that the UV-sensitive targets are nucleic acids. Ribonucleoprotein complexes as they are present in the pole plasm have not been identified in the periplasm of the egg of *Drosophila* (MAHOWALD, 1968), although OKADA and WADDINGTON (1959) report the occurrence of several types of granules. By puncturing early cleavage stages of *Drosophila*, ILLMENSEE (1972) could induce imaginal defects that were correlated with the site of damage. Further evidence that the periplasm exerts a determinative influence upon the immigrating nuclei comes from the mutant *bicaudal* of *Drosophila melanogaster*. This recessive genetic syndrome causes the formation of a second abdomen symmetrically located in the anterior egg region (BULL, 1966). The transformation of anterior qualities into posterior ones is maternally inherited and independent of a normal paternal genome. This fact indicates that determination, at least for a gross anterior-posterior organization of the embryo, is laid down in the cytoplasm during oogenesis.

In summary, there are several lines of evidence for the concept that basic processes of determination occur at the time when omnipotent cleavage energids enter the egg's periplasm.

C. Number of Primitive Cells

What is the number of primitive cells (progenitor cells) that form a disk's initial cell population? An answer to this question is difficult to obtain if conventional histological techniques are applied. Best analyzed so far is the genital disk of *Drosophila melanogaster* (LAUGÉ, 1969a, b). It first appears in histological sections as a thickening, consisting of only two (or very few) cells, in the ventral epidermis of the late embryo. Sections through later stages show more and more cells in the anlage. These histological sequences do not tell us whether the initial two cells divide and form the entire anlage, or whether the growing disk is continuously supplied by additional cells from the surrounding presumptive epidermis. Such a supply throughout the larval instars of *Drosophila virilis* has been suggested by NEWBY (1942). He arrived at this conclusion because mitoses appeared to be very rare in the disk anlage throughout the larval growth period. In her recent study, BABCOCK (1971) states, although without presenting experimental evidence, that "during larval life the (female) genital disk increases in size through the addition (where from?) of new cells". On the other hand, she observed mitotic figures in the disk's inner epithelium which gives rise to the chitinous part of the genitalia. This indicates "that the increase in cell number is, at least in part, the result of the division of primordial disk cells." Observations made with gynandromorphs provide strong evidence against a continuous supply of cells. GARCIA-BELLIDO and MERRIAM (1969a) consider it possible that the genital disk arises from only two initial cells: the external genitalia of gynandromorphs were either of one sex only (308/379 cases) or approximately 50% male and 50% female (71/379 cases) in one contiguous patch. Addition of new epidermal cells which could either be male (marked with $y=yellow$) or female (wildtype color) would lead to a checkerboard-kind of mosaicism instead of only two contiguous patches. The absence of checkerboard mosaicism strongly supports the idea that once the initial cells of a disk are segregated, the anlage becomes a closed system, at least for the presumptive chitinous structures, without any further supply of cells from outside. One would expect that such a supply from larval epidermal cells should necessarily come to a halt after these start to become polytene. Unfortunately, the exact stage at which this process begins is not known. In *Calliphora*, significant differences in the diameter of nuclei between the larval epidermal cells and the imaginal cells of the abdomen are already manifest in the first larval instar (BAUTZ, 1971). On the other hand, it is conceivable that the presumptive mesodermal cells (mesenchymal = adepithelial cells, see POODRY and SCHNEIDERMAN, 1970) could enter the disks sometime later during larval development (see EL SHATOURY, 1955).

A more elegant and also more accurate approach to the problem was first undertaken by STURTEVANT (1929) and later led to perfection by STERN (1940). The method uses gynandromorphs which, as outlined on p. 5, consist of two genetically differing populations of cells. STURTEVANT noticed that some disks were more frequently composed of the two genotypes than other disks. His interpretation was that the frequency with which a disk is mosaic will be the higher the more cells are initially included in the primitive disk anlage. To estimate this number, the clones must have been initiated prior to the determination of the disks. This is most likely the case since the loss of one of the X chromosomes occurs in the first or one of the first zygotic divisions (discussed by POSTLETHWAIT and SCHNEIDERMAN, 1971a). We furthermore assume that in the course of development each primitive cell of a disk undergoes the same average number of divisions (see, however, p. 17). Under these conditions, the reciprocal of the smallest fraction which can be genetically XX or XO in a mosaic disk yields an estimate of the number of progenitor cells. This number has recently been determined for all major disks and found to be around 10—40 (see Table 1). A method which is less dependent on equal division rates in a disk but requires a large number of gynandromorphs was used by POSTLETHWAIT and SCHNEIDERMAN (1971a) to calculate the number of primitive cells forming the antenna. Their estimate falls well within the range of values obtained with the first technique (Table 1).

The gynandromorph method, of course, is applicable only to those derivatives of a disk for which autonomous, X-linked cell marker mutants are available. So far, this is not yet the case for adepithelial cells which are included in some disks (see URSPRUNG, this volume) and which later give rise to mesodermal structures such as the muscles of the leg (LÄUBLI and URSPRUNG, unpubl.), or of the genital apparatus (HADORN, BERTANI and GALLERA, 1949; URSPRUNG, 1959). It should also be emphasized that the gynandromorph data do not tell us anything about the time of primary disk determination.

Table 1. Number of primitive cells *(n)*

Disk or primordium	n	Ref.
Antennal	7—9	Postlethwait and Schneiderman (1971)
Eye anlage	2	Becker (1957)
Eye-antennal	13	Garcia-Bellido and Merriam (1969a)
Ventral thoracic (leg)	~20	Garcia-Bellido and Merriam (1969a); Bryant and Schneiderman (1969)
Dorsal mesothoracic (wing)	16	Stern (1940)
	12	Garcia-Bellido and Merriam (1969a)
	11	Bryant (1970)
	~40	Ripoll (1972)
Abdominal histoblasts	8	Garcia-Bellido and Merriam (1971b)
Genital	~ 2	Garcia-Bellido and Merriam (1969a)

D. Role of Disks in Larval Life

When a disk is extirpated from a larva, development can nevertheless proceed, and an adult fly can emerge (see p. 21). This result indicates that the disk primordia are non-functional and hence dispensable structures during larval life. This view is now further substantiated by an analysis of induced lethal factors which cause death after the larval stages (Shearn et al., 1971). 18 new mutants were isolated in which either all (group A) or some (groups B and E) disks are missing. The larval development was essentially normal in all cases.

V. Growth and Morphogenesis

The preceding section has introduced the imaginal disk as a closed system of a few ectodermal cells that are set aside most probably around the time of blastoderm formation and whose clonal descendants will give rise to all of the ectodermal imaginal cell types differentiated by a particular disk.

A. Dynamics of Growth

For the analysis of the dynamics of growth in imaginal disks, histological techniques, autoradiography, and induced mitotic recombination have been used as experimental tools. The following section will attempt to give a short survey of the relevant facts and open questions in this aspect of disk development.

1. Histological Analysis

Cell counts at different times throughout larval development have been made by Chevais (1944) and Becker (1957) for the eye-antennal disk of *Drosophila melanogaster*. According to Becker (1957), this primordium consists of about 130 cells at the beginning of the second larval instar. It increases exponentially to some 900 cells shortly before the end of this instar. The rate of proliferation slows down just prior to the second larval molt when about 1300 cells are present. This corresponds to an increase in cell number by a factor of 10 in 21 h and represents an average cell cycle length of ca. 6 h $\left(21 \text{ h divided by } \frac{\log 10}{\log 2}\right)$. The cell number then increases again during the third larval instar and reaches a final number of some 30000 cells in the mature disk. A cytological examination revealed three "mitotic waves" during the second larval instar, with the mitotic index varying between 2% (minimum) and 8% (maximum). This phenomenon points to a partial synchronization of cell proliferation.

As TIMM (1970) has shown for the regenerating wing disk of *Ephestia kühniella*, the mitotic index is not a reliable indicator of mitotic activity. In this disk, changes in growth rate, as measured by the increase in cell number per unit of time, are not accompanied by corresponding changes in the mitotic index. Rather, changes in growth rate appear to be correlated with changes in the length of the cell cycle in such a way that high mitotic activity is achieved by shortening the cell cycle period. An autoradiographic analysis of the same system (LÖBBECKE, 1969) indicated that the duration of the S-phase decreases drastically from about 12 h to about 6 h during the last larval instar. Very likely, the cells in the wing disk are not homogeneous with respect to mitotic activity. In fact, MUTH (1961) has reported that mitoses are locally restricted on the wing disk of *Ephestia* at certain stages of development.

2. Induced Mitotic Recombination

The analysis of the dynamics of growth in imaginal disks of *Drosophila* has been elegantly carried out with the technique of mitotic recombination. This process can

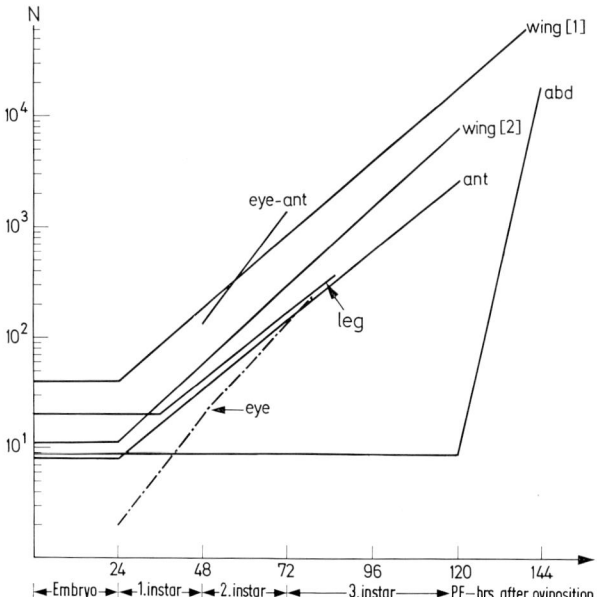

Fig. 3. Growth curves of various imaginal disks of *Drosophila*. N is the number of cells estimated from genetic mosaics (gynandromorphs and mitotic recombination, see p. 6), except for *eye-ant* which is based on cell counts in squash preparations. *abd*, abdominal histoblasts (GARCIA-BELLIDO and MERRIAM, 1971b); *ant*, antennal disk (POSTLETHWAIT and SCHNEIDERMAN, 1971); *eye*, presumptive cells of the eye region (BECKER, 1957); *eye-ant*, eye-antennal disk (BECKER, 1957); *leg* leg, disk (BRYANT and SCHNEIDERMAN, 1969); *wing*, wing disk ([1]: GARCIA-BELLIDO and MERRIAM, 1971a; RIPOLL, 1972; [2]: BRYANT, 1970) *PF* puparium formation

be induced at any desired time during disk development (see p. 5). Data are now at hand for all major disks (see Fig. 3). It is apparent that all imaginal cells remain stationary during embryogenesis, but then grow continuously and exponentially throughout the larval period. The doubling time, which is an estimate for the average length of the cell cycle, varies between 6 and 15 h for different disks. The abdominal histoblasts represent an exception inasmuch as their cell number does not appear to increase until the time of puparium formation when they start to divide at the rapid rate of approximately one cell cycle every 2—3 h. The differences just described cannot be generalized for other Diptera: in *Dacus tryoni* (ANDERSON, 1964), in *Calliphora erythrocephala* (DAHLHEIM, 1967; BAUTZ, 1971), and in *Musca domestica* (unpublished data, obtained with induced mitotic recombination), the abdominal histoblasts grow steadily by cell multiplication during the second half of the larval period.

3. Regional Differences in Mitotic Activity

Whereas the curves in Fig. 3 indicate that the disks represent homogeneous cell populations, a more detailed analysis reveals that this is not the case. Clone sizes vary considerably even when two clones are located within the same eye (BECKER, 1957), leg (BRYANT and SCHNEIDERMAN, 1969) or wing of an individual (BRYANT, 1970; GARCIA-BELLIDO and MERRIAM, 1971a). Furthermore, cell numbers in clones on the wing deviate significantly from 2^n-values indicating that a part of a clone may stop dividing earlier than other cells of the same clone. More important are non-random, reproducible differences between subpopulations of cells of a disk. BECKER (1957) has shown that among the twin spots induced in the first larval instar the clone in the anterior eye region consistently exceeded its twin partner in the posterior eye region by one to two cell divisions (Fig. 2). Since in twin spots, variables such as variation in age or X-ray damage will equally affect both daughter cells resulting from mitotic recombination, the difference in clone size reflects regionally controlled differential mitotic activity within a disk.

Similar observations have been made for the wing disk (GARCIA-BELLIDO and MERRIAM, 1971a) where clones in the anterior part of the wing are about half the size of clones in the posterior part, and for the antennal disk (POSTLETHWAIT and SCHNEIDERMAN, 1971a) where mitotic activity was found to be higher in the ventral than in the dorsal part of the third antennal segment. Data from a clonal analysis of regeneration in half genital disks of *Drosophila* indicate that the regenerated "new" side of the disk originated from a small number of cells which divided more frequently than the cells forming the "old" half (ULRICH, 1971; NÖTHIGER and ULRICH, in prep., see also Fig. 5b, c). Such non-random regional differences are indicative of control mechanisms, and furthermore may have a bearing on determination and morphogenesis (see p. 17).

The fact that the cells of imaginal disks consist of dividing and non-dividing subpopulations has interesting implications for problems of growth regulation and differentiation. Cells might be "stored" in a phase of the cell cycle from where they can be called upon to assume their specific developmental roles. CAMERON and CLEFFMANN (1964) showed that the G_2-phase can easily be prolonged in starved chickens, and LÖBBECKE (1969) found the same phase to be variable in the wing disk of *Ephestia*. In the non-dividing cells of the abdominal histoblasts of *Drosophila*, mitotic recombination which requires four chromatids to be present can be induced throughout the larval stages (see Fig. 3). This suggests that these cells are arrested in the G_2-phase or early prophase. PENTZ and KRAUSE (1968) were able to initiate a diapause in *Plodia interpunctella* (*Pyralidae*, Lepidoptera) which resulted in all cells of the wing imaginal disks being stalled in an early prophase. In these cells mitotic activity and further development could later be induced by higher temperature. If such an inducer has local access only, differential growth may result.

4. Criticism

The conclusions drawn in this section should not be accepted without criticism. Most of the limitations adhering to the technique of induced mitotic recombination for this kind of analysis were already listed on p. 7. In addition, the phenomenon of programmed cell death in particular regions of a disk as reported by SPREIJ (1971) and FRISTROM (1972) introduces a new source of error. Marked clones in such a region would consistently be smaller. This, in turn, would lead to an underestimate of the number of cell divisions and consequently to an overestimate of the length of a cell cycle.

B. Morphogenesis

A main feature of morphogenesis in imaginal disks is folding. A disk makes its first histological appearance through invagination of the prospective disk epithelium during embryogenesis. While growing during the larval stages the epithelium becomes thrown into a complex folding pattern which is characteristic for each disk (AUERBACH, 1936). During metamorphosis, when the disks are everted, or evaginated, the initial invagination process is essentially reversed. It has been studied *in situ* by WEHMAN (1969) and POODRY and SCHNEIDERMAN (1970), and *in vitro* by MANDARON (1970, 1971) and FRISTROM (see review, this volume). The following six phenomena are relevant in connection with disk morphogenesis:

a) *Active migration, or passive movement of cells or tissues.*
b) *Orientation of the mitotic spindle apparatus.*
c) *Local differences in mitotic activity.*
d) *Local differences in cell size and shape.*
e) *Localized cell death.*

Our modest knowledge about this key problem of development is mainly based on analyses of genetic mosaics.

a) *Active cell migration*, as it is described for neural crest cells of vertebrates (see TRINKAUS, 1966, for review), has so far not been demonstrated to play an important role in imaginal disk development. Marked clones in gynandromorphs form contiguous areas, with cells of different genotypes being intermingled only rarely and to a slight degree (STURTEVANT, 1929; GARCIA-BELLIDO and MERRIAM, 1969a). BRYANT (1970) who analyzed genetically marked clones on the thorax observed rare displacements of cells outside of a clone. Similar observations were made with clones on the wing (GARCIA-BELLIDO and MERRIAM, 1971a) and on the abdomen (GARCIA-BELLIDO and MERRIAM, 1971b). However, this occasional splitting of a clone does not necessarily indicate developmentally significant active cell migration. Rather, it appears to be due to a passive displacement of cells following mitotic divisions. If such a mechanism were to play a role in morphogenesis it ought to exhibit a regional regularity. This is not the case, and the occurrence of displaced cells appears rather capricious.

A regular movement of a whole group of cells takes place in the primordia of the male basitarsus (TOKUNAGA, 1962; see GARCIA-BELLIDO, this volume), and of the antenna (POSTLETHWAIT and SCHNEIDERMAN, 1971a). For the antenna, the authors noticed that a clone initiated so early in development as to extend over the whole antenna was discontinuous between the second and third antennal segment whereby the two segments appeared to have rotated relative to each other by about a quarter of a turn. When antennal disks were dissected from late third instar larvae and early pupae, the shift of the third antennal segment, inferred from the morphology of the clones, could be directly observed. The movement of the prospective sex comb teeth, on the other hand, occurs within one segment, the basitarsus, and almost necessarily involves a shift of cells within an epithelium (see Fig. 7 in GARCIA-BELLIDO, this volume). No experimental data are available concerning the mechanics of these movements. The observations leave open the question whether active migration or passive shift occurred.

b) *Orientation of the mitotic spindle apparatus.* Indications that cell divisions might be oriented within the single-layered epithelium of the growing imaginal disks of *Drosophila* came from analyses of gynandromorphs. If a wing (GARCIA-BELLIDO and MERRIAM, 1969b; RIPOLL, 1972), a leg (BRYANT and SCHNEIDERMAN, 1969), or an antenna (POSTLETHWAIT and SCHNEIDERMAN, 1971a) were mosaic, the borderlines ran frequently parallel to the longitudinal axis of these appendages. The same observations were made with clones which were induced by mitotic recombination early in development. The best analysis on the non-random shape as well as location of marked clones is available for the eye (BECKER, 1957; BAKER, 1967). BECKER could identify 8 sectors in the lower part of the eye. The borders of genetically marked sectors ran preferentially in three main directions: horizontal, vertical, or from dorsoposterior to ventroanterior (Fig. 2). Individual clones occurred either as stripes or as wedge-shaped sectors. Essentially the same pattern of cell-lineage was found by BAKER (1967) for *Drosophila virilis;* horizontal stripes as well as triangular sectors were also detected in the eye of *Musca domestica* (NÖTHIGER and DÜBENDORFER, 1971).

Twin spots (Fig. 2) provide information on the orientation of the division which gave rise to the two daughter clones. BECKER (1957) showed for the eye that this orientation is non-random. The preferential directions are characteristic for a given stage of development, which suggests programmed systematic changes in the orientation of the mitotic spindle in relation to other developmental events. In principle, the same statement holds for the leg disk (BRYANT and SCHNEIDERMAN, 1969), for the wing disk (BRYANT, 1970; GARCIA-BELLIDO and MERRIAM, 1971a), for the antennal disk (POSTLETHWAIT and SCHNEIDERMAN, 1971a), and for the abdominal histoblasts (GARCIA-BELLIDO and MERRIAM, 1971b). Marked clones on the abdominal tergites of *Musca domestica* also grow preferentially in a medio-lateral direction (unpublished data).

The inference from genetic mosaics, that cell divisions can be oriented, is confirmed by a cytological study of the wing disk of *Drosophila* (STUMPF, 1956). At 17 h after puparium formation the mitotic spindles were preferentially at a right angle, and at 22 h preferentially parallel to the long axis of the presumptive wing.

Whereas the clonal analysis suggests that the mitotic spindles may be oriented, and such an orientation can explain the observed shapes of the clones, the data are contributing nothing to an understanding of the mechanisms which control this phenomenon. Also unexplained is the amazing precision with which a clone comes to stop at the appropriate place of the wing margin, so that the typical circumference of the wing is formed (GARCIA-BELLIDO and MERRIAM, 1971a). No experiments have been devised so far to attack this obviously difficult and fascinating problem.

c) *Regionally patterned differences in mitotic activity* are a common phenomenon revealed by genetically marked clones which were induced early in larval development (see p. 14). One can imagine that localized differences in the rate or number of cell divisions may cause a bending, folding, or bulging of an epithelium. POSTLETHWAIT and SCHNEIDERMAN (1971a) are assigning a morphogenetic role to the elevated mitotic rate in the ventral region of the antennal disk of *Drosophila*. It would be extremely difficult to experimentally demonstrate a morphogenetic significance for this process, e.g. by causing an alteration in the regional pattern of cell divisions which would then lead to a predictable morphological change.

d) *Regionally patterned differences in cell size or shape* to which a morphogenetic role has been assigned, were found on the third antennal segment (POSTLETHWAIT and SCHNEIDERMAN, 1971a). Cells in the ventral region are more than twice as large in surface area as cells in the dorsal region. In the authors' view this indicates that the bend in the third antennal segment is at least partially due to localized differences in cell size or shape.

The most dramatic morphogenetic processes take place when the disks are everted during metamorphosis. Within a few hours, the complexly folded disk is evaginated, and gross changes in the shape of the disk occur. POODRY and SCHNEIDERMAN (1971) discovered that a mild treatment *in vitro* with trypsin, pronase, or collagenase could mimic the initial steps of prepupal morphogenesis in leg and wing disks of the late third instar *(Drosophila)*. The primordia of the tarsus and of the other leg segments unfolded after 5—10 min in 0.1% trypsin in very much the same way as they do normally *in situ* at the onset of metamorphosis. Histological sections revealed that the cells changed from a tall columnar to a squamous shape. An electron microscopic analysis showed that all types of intercellular junctions as well as the microtubules and other cell constituents remained unchanged even after 4 h of trypsinization. The authors hypothesized that a change in cell shape, brought about by a change in intercellular adhesiveness (GUSTAFSON and WOLPERT, 1967), is the main factor involved in prepupal disk morphogenesis, and that in this process the microtubules which have been assigned morphogenetic roles in other systems (e.g. KARFUNKEL, 1971), appear to be without significance. However, the situation is apparently more complex (see FRISTROM, this volume).

e) *Localized cell death* appears to be a main factor in the morphogenesis of the vertebrate limb (SAUNDERS and FALLON, 1966). It also occurs in imaginal disks of *Drosophila* and *Calliphora* (FRISTROM, 1969, 1972; SPREIJ, 1971). SPREIJ found that the appearance of dead cells in the various imaginal disks (eye-antenna, leg, wing, haltere) exhibited a spatial and temporal pattern which was similar for *Calliphora* and *Drosophila*. The first groups of degenerating cells became recognizable only after the second larval moult. A constant regional pattern of nests of degenerating cells emerged later in the third instar. As long as the disk epithelium is not yet invaginated (early third instar) dead cells are found in groups along the periphery and randomly distributed in the center of the disk. When invagination of the epithelium takes place the number of degenerating cells decreases in the center. Along the periphery, the groups are found at the base of the prospective appendages (leg, wing, etc.). With further outgrowth in the prepupal period the number of degenerating cells increases again in the invaginated part of the appendages. Here, they are now also located in distinct areas.

An analysis of several wing and eye mutants of *Drosophila melanogaster* (FRISTROM, 1969, 1972; SPREIJ, 1971) showed that the particular shape of the adult organ in these mutants could be correlated with a high degree of cell death occurring in specific regions of the disks.

Based on these observations, SPREIJ (1971) suggested that localized cell death may have a morphogenetic significance in the development of imaginal disks. Several questions, critically discussed by the author, are still open. He proposed a model which accounts for the typical folding and invagination processes in imaginal disks by regionally patterned cell death. By comparing his results with data which indicate oriented cell divisions (see p. 16) the author concluded that localized cell death is just one factor in disk morphogenesis.

f) *Trying to arrive at a final statement* we will have to keep in mind that all conclusions from the preceding section rely on indirect evidence. But it seems safe to say, that no one of the factors discussed can be assigned the sole role in the complex morphogenetic processes of moulding the disks. We are led to believe that shape is achieved by a cooperation of several of these factors.

As a last question, we may ask what is the relationship between the morphology of a disk and its prospective developmental fate. WADDINGTON (1962) has proposed that the specific folding pattern of a disk causes its determination for specific adult structures. His conclusion was based on observations with mutations which changed the folding pattern and (hence?) the determination of the mutant disk. In my view, more recent experiments do not support the hypothesis. When disks are fragmented, or even dissociated and reaggregated, their folding pattern is grossly disturbed. Nevertheless, the cells maintain their specific state of determination (see p. 22). This is best demonstrated by "faulty mosaics" (NÖTHIGER, 1964). When a few cells of one disk were trapped in a foreign environment of another disk, they differentiated autonomously according to their origin. One might still object that the folding conveyed a specific determination on the disks, but that once this is established, the cells are capable of autonomous differentiation. Recent experiments (HADORN et al., 1968; CHAN and GEHRING, 1971) in which dissociated embryonic cells were cultured in adult females suggest that irregularly folded imaginal disk tissues nevertheless acquired a specific state of determination. I, therefore, believe that the folding pattern may reflect rather than cause the state of determination.

VI. The Switch from Cell Multiplication to Cell Differentiation
A. Competence

It is generally accepted that in holometabolous insects adult differentiation of the imaginal disks is dependent on hormonal conditions. Metamorphosis in insects is initiated by a high level of ecdysone in connection with a low level of juvenile hormone

(GILBERT and SCHNEIDERMAN, 1964; OBERLANDER, this volume). In adult flies where no detectable ecdysone is present (SHAAYA and KARLSON, 1965, for *Calliphora*) the cells of imaginal disks of *Drosophila* can multiply indefinitely; their imaginal differentiation is induced whenever they are subjected to conditions of metamorphosis by transplantation into larval hosts (HADORN, 1966). However, hormones are not the only parameters for the initiation of adult differentiation. Eye-disks of *Drosophila*, taken from larvae some 50 h after oviposition and transplanted into mature larval hosts, exhibited only partial and poor differentiation (BODENSTEIN, 1939). This observation, later confirmed by HADORN, BERTANI, and GALLERA (1949) for the male genital disk, indicates that an imaginal disk has to be competent to react to the humoral stimuli. This state of responsiveness appears to be reached around the second larval moult (BODENSTEIN, 1939; GATEFF, 1971; MINDEK, 1972), i.e. coincident with the so-called "seventy-hour change" (BEADLE, TATUM, and CLANCY, 1938). These authors noticed that larvae removed from food prior to 70 h after oviposition and placed without food on moist filter paper continued to live for several days, but finally died without ever exhibiting any signs of imaginal differentiation. Larvae similarly removed from food shortly after 70 h failed to increase in size, but formed puparia approximately simultaneously with normally fed controls. BODENSTEIN (1939) concluded that important developmental changes must occur in the eye-disk at this stage. Disks from larvae prior to 70 h and transplanted into mature larvae were not able to differentiate normally, whereas eye-disks taken from larvae after this time differentiated normal eye structures. It appears as if the competence to react to the conditions of metamorphosis is conveyed to the disks at the time of the "seventy-hour change". There are only minor differences in time between the various disks, although the different prospective regions of a disk do not reach competence simultaneously, but in a fixed temporal sequence (GATEFF, 1971).

The response of a disk is not an "all-or-none reaction". When competent eye disks (SCHLÄPFER, 1963) or male genital disks (NÖTHIGER and OBERLANDER, 1967) were subjected to suboptimal hormonal conditions in adult female hosts, only eye pigments or contractile proteins were synthesized but no other imaginal structures differentiated.

B. The Acquisition of Competence

1. *Humoral Factors*

As first demonstrated by HADORN et al. (1968) presumptive imaginal cells of *Drosophila* embryos can also acquire competence in the humoral milieu of adult females. This fact has been utilized to culture and bring to maturity embryonic cells of nuclear transplants (ILLMENSEE, 1970; SCHUBIGER and SCHNEIDERMAN, 1971), or dissociated blastoderm stages (CHAN and GEHRING, 1971). Now SCHNEIDER (1972) was able to culture embryonic material *in vitro* in a complex medium and to obtain growing cells which are competent to differentiate imaginal structures upon transplantation into mature larvae. As MINDEK (1972) showed this process of maturation can even proceed in a metamorphosing pupal host. Eye-antennal disks of freshly hatched larvae were transplanted into larvae ready to pupate. After metamorphosis of the host, the transplants had developed into typical large eye-antennal disks without any signs of imaginal differentiation. When such disks were reinjected for a second time into mature larvae they produced the full and normal inventory of eye and head structures.

These experiments reveal an impressing autonomy in the development of imaginal disks: the acquisition of competence is not only completely independent of the rhythmical hormonal changes which occur in a growing larva, but neither can it be disturbed by ecdysone levels which cause other disks in the same individual to undergo imaginal differentiation.

2. Mutations Affecting Competence

One might reasonably expect that mutations could interfere with the process of acquiring competence. Mutations that lead to a complete lack of response would necessarily be lethal in the pupal phase. Unfortunately, this field has not been given any attention in the past. Only recently, SHEARN et al. (1971) have accumulated a number of lethals some of which interfere with development of imaginal disks. Their groups B and C, in which extremely small disks are unable to metamorphose normally, could represent mutations partially blocking competence. STEWART, MURPHY, and FRISTROM (1972) have performed a similar analysis. They induced a number of X-linked pupal lethals in *Drosophila*. In two of these mutants imaginal disks developed that were indistinguishable from normal wildtype disks, as judged from dissection. However, when transplanted into wildtype larval hosts the mutant disks either failed completely to differentiate any adult structures, or did so only very poorly. In animals homozygous for *lethal-2 giant larvae (l(2) gl^4)* the imaginal disks grow to a near-normal size but are very disorganized and unable to enter adult differentiation. Concomitantly with the loss of this capacity the cells of the imaginal disks are transformed into a non-invasive neoplasm. They kill their hosts within 7—14 d when transplanted into adult flies (GATEFF and SCHNEIDERMAN, 1969). It is not known whether the rapidly growing cells of the neoplasm are ectodermal in origin, or whether they derive from the adepithelial cells in the disk.

Some of HADORN's (1966) cell lines have become what he called "atelotypic" ("C-type" of HADORN and GARCIA-BELLIDO, 1964); they grow rapidly in adult hosts, but do not metamorphose any more when transplanted into larval hosts. Because of the low frequency with which such changes occurred they were assumed to represent somatic mutations. Now GATEFF (pers. comm.) claimed that the loss of competence is not irreversible. When atelotypic cells of *D. melanogaster* were cultured in adult females of *D. virilis* they grew much more slowly, and upon transplantation into larvae of *D. melanogaster* were now able to differentiate imaginal structures.

3. Causal Analysis

Competence may simply be a function of the number of cells present; or of the time having passed between the singling-out of a disk and the onset of metamorphosis; or it may require a series of complex metabolic processes. The first possibility is experimentally ruled out for *Drosophila*. Even very few cells are able to differentiate, as concluded from the frequent occurrence of isolated bristles after dissociation (see e.g. NÖTHIGER, 1964; GARCIA-BELLIDO, 1966). We (MINDEK and NÖTHIGER, in prep.) have also tested the second variable (time): incompetent disks from 53 h old larvae were cultured for 2 d in adult females kept on a protein-free diet. This prevented the imaginal cells from multiplication (GARCIA-BELLIDO, 1967). The disks remained small, and upon injection into mature larvae were still incompetent to respond to the hormonal stimuli of metamorphosis. In controls kept on standard yeast-food the disks grew to several times their initial size, and when transplanted into mature larvae differentiated the full set of prospective imaginal structures. The block caused by the protein-free diet is not irreversible. If, after this treatment, the flies were transferred for another 2 d on complete food, the disks resumed growth and when injected into mature larvae could now respond by normal differentiation. The experiment eliminates time as a parameter for aquiring competence. It leaves open the question whether simply a number of cell divisions, or other metabolic processes are required for the acquisition of competence and hence of the capability of the cells to switch from multiplication to imaginal differentiation.

VII. The Acquisition and Propagation of a Given State of Determination

The following section concerning the problem of determination in imaginal disks will be presented in the form of six short statements.

A. Individual Disks and Parts of Disks are Determined to Develop into Specific Regions of the Adult Fly

When whole imaginal disks are isolated and transplanted into larval hosts they invariably develop according to their prospective fate. The inventory of differentiated adult structures corresponds to what is formed *in situ* (references in SCHUBIGER, 1968). Conversely, if a disk is extirpated from a larva, the emerging adult is deficient for a precisely corresponding region (ZALOKAR, 1943; MURPHY, 1967; WILDERMUTH, 1968; DÜBENDORFER, 1971). This autonomy of development shows that under the conditions of transplantation, an imaginal disk represents a rigidly determined primordium (see, however, p. 28).

Disks can also be mechanically divided into a number of fragments. Such fragments, transplanted into late third instar larvae, only differentiate into parts of a disk's normal adult derivatives. Conversely, elimination of small areas from a disk results in an incomplete inventory of differentiated adult structures. It is concluded that in mature larvae, a disk represents a conglomeration of areas in which the cells are determined to develop into specific structures differentiated by that disk. This mosaic developmental behavior allowed the construction of fate maps (see e.g. SCHUBIGER, 1968). Such fate maps have been established for all major disks of *Drosophila melanogaster* (see Table 3, in GEHRING and NÖTHIGER, 1972).

B. In Late Third Instar Larvae, Even Single Cells are Determined to Develop into the Cell Types of a Specific Region of the Adult Fly

The fragmentation experiments can give no clues as to whether an individual cell of an anlage is already programmed for its particular pathway of development, or whether the state of determination is carried and maintained by populations of cells, or fields (WADDINGTON, 1966). This problem may be approached by dissociating and reaggregating the cells of two genetically marked disks, a procedure introduced by HADORN, ANDERS, and URSPRUNG (1959).

The general observation is that reaggregates can form integrated mosaics when the cells in the mixture are derived from the same type of disk (homonomous combination; e.g. leg × leg), but do not when the cells are derived from different types of disks (heteronomous combination; e.g. leg × wing). When cells from two different regions on the fate map of a disk are mixed, mosaics do not form either, except for those structures whose presumptive cells are situated in the border area between the two regions (GARCIA-BELLIDO, 1966; TOBLER, 1966). This absence of mosaicism supposedly results from a sorting-out of cells which differ in the quality of their state of determination (heterotypic cells) whereas mosaics may be formed as a consequence of random collisions between cells carrying identical qualities of determination (isotypic cells). The reader will find a more thorough evaluation of these phenomena in the chapter by GARCIA-BELLIDO (this volume). Sometimes, "faulty mosaics" are found: these consist of single cells, or very small groups of cells, from one type of disk being trapped among a large number of cells from another type of disk. The trapped cells differentiate autonomously according to the determination of their disk of origin (NÖTHIGER, 1964). From this, of course, we cannot conclude that these cells were specifically determined to develop into one particular type of bristle; but the finding clearly shows that no "assimilative induction" occurred (see below, and GARCIA-BELLIDO, this volume). These experiments led to the idea that the individual cell rather than a blastema is the carrier of determination.

This concept has recently been challenged by POODRY, BRYANT, and SCHNEIDERMAN (1971). They have mixed dissociated leg disks of two different genotypes ($y\,sn^3$ and mwh in a ratio of 1:1) in a background of varying amounts of heterotypic dissociated wing disk cells (sn^3; e^{11}). Whereas cells from leg and wing disks did not form any integrated mosaics, the cells from leg disks were capable of cooperating in mosaic leg structures. The frequency of these mosaics decreased with increasing amounts of wing disk cells. The authors concluded that a "directed migration of individual cells which would take up their preassigned specific positions in the new pattern" does not occur. — To my knowledge, the hypothesis in quotes has never been proposed, and furthermore, I think the data would not eliminate it; the intermingled wing cells will increase the average distance between two isotypic cells whose velocity of migration in a reaggregated cell clump is not known. We have no reason to assume *a priori* that the time allowed to pass between reaggregation and differentiation would in all cases be sufficient for isotypic cells to overcome the increasing distances. POODRY, BRYANT, and SCHNEIDERMAN (1971) furthermore suggest that "a considerable amount of repatterning occurs among the randomly associated cells". The term "repatterning" has not been defined but probably implies a "reprogramming", or a new determination of cells. Again, in my opinion, the data do not show that the dissociated cells are actually assigned new developmental roles, since the experimental design did not allow to detect cases of "assimilative induction", i.e. cells that were obviously reprogrammed under the influence of heterotypic cells. Furthermore, if most cells of imaginal disks are in a "prefinal" state of determination at the end of the larval stages (see GEHRING and NÖTHIGER, 1972; and p. 23), there is no need to invoke a "repatterning" to explain cases of apparently normal patterns after dissociation and reaggregation.

C. Determination Proceeds Stepwise

Genetically marked patches of gynandromorphs extend over large areas of the fly's surface comprising the derivatives of several disks (GARCIA-BELLIDO and MERRIAM, 1969a). This indicates that the different disks became determined considerably after the loss of one of the X chromosomes (see p. 5). Analyses of marked clones which were induced at successively later times up till metamorphosis, revealed that the developmental capacities of imaginal disk cells become progressively restricted (BECKER, 1957; BRYANT and SCHNEIDERMAN, 1969; BRYANT, 1970; GARCIA-BELLIDO and MERRIAM, 1971a; POSTLETHWAIT and SCHNEIDERMAN, 1971a). Clones induced in the wing disk between 2 and 24 h after oviposition were found to comprise both the upper and lower surfaces of the wing blade, or the upper wing surface and the mesonotum. A segregation into three cell populations became apparent when larvae were irradiated at 48 h after oviposition; among a number of large clones, 21 were found that were confined either to the upper, or to the lower wing surface, or to the mesonotum (BRYANT, 1970). Marginal clones invariably "stopped" at the edge of the wing, and frequently ran parallel to it for hundreds of hairs without ever extending on both wing surfaces (GARCIA-BELLIDO and MERRIAM, 1971a). This suggests determinative events leading to the establishment of these three populations of cells in the wing disk of the late first or early second instar. In the second half of the third larval instar, further steps of determination are apparently separating the presumptive bristle cells of the wing margin from the presumptive hair cells of the wing blade (GARCIA-BELLIDO and MERRIAM, 1971a). The technique of mitotic recombination is unable to trace later developmental restrictions *in situ:* clones will eventually become so small that they cannot be expected to comprise two or more different structures. But experiments with dissociated disks point to even more specific states of determination at the onset of metamorphosis (see GARCIA-BELLIDO, this volume).

The determination of bristle-forming and hairforming cells does not take place at the same time in all disks. In the abdominal histoblasts e.g. it occurs only some hours after puparium formation (GARCIA-BELLIDO and MERRIAM, 1971b). The histoblasts, however, behave very differently from other disks in several respects (see p. 1 and Fig. 3). It therefore seems more reasonable that the time at which a developmental decision is made, is related to the number of cell divisions rather than time still to be passed until imaginal differentiation.

D. Determination has Reached a "Prefinal" State at the End of Larval Development

It is difficult to assess with certainty the exact state of determination of an individual cell in a disk of a mature larva. This is because the cells are eluding our analysis: as of now, we can only identify their state of determination *after* they have passed through metamorphosis, and we have no ways yet to know what their exact developmental status was at the beginning of the experiment. We will now present some findings which suggest that the cells may be in a "prefinal" state of determination at the end of the larval period. By "prefinal" we mean that the cells are no doubt committed to a very narrow pathway of development, e.g. to form a certain region within the basitarsus, but not yet to one particular cell type within this region (bristle or trichome, type of bristle). All experiments were carried out with disks from late third instar larvae.

The first case is provided by the "bracts", special forms of trichomes, which arise under the inductive influence of a bristle in certain regions of the leg and wing. Upon dissociation bracts are unable to differentiate in isolation, indicating (see p. 4) that they were not yet definitively determined in the disk (TOBLER, 1966; GARCIA-BELLIDO, 1966, and this volume). — The male anal plates of *Drosophila séguyi* are equipped with four different types of bristles (HADORN, 1953). Experiments aimed at localizing these bristles in a fate map failed. Only the anal plate primordium as a whole could be mapped (LÜÖND, 1961). It appears that the cells of this primordium carry only the general determination for anal plates, and that the specific roles for the four bristle types have yet to be assigned to the cells.

Further arguments are based on cell lineage analyses in duplicating disk fragments (see p. 28). We have initiated such an investigation several years ago in our laboratory (NÖTHIGER and ULRICH, in prep.). Medially halved male genital disks were allowed a limited amount of cell proliferation in order to duplicate themselves and to form a complete symmetrical genital apparatus (URSPRUNG, 1962; NÖTHIGER and SCHUBIGER, 1966). The experimental design was to induce a clone by mitotic recombination just before the process of duplication had begun. The initial half disks were oriented so that after duplication an "old" and a "new" side of the symmetrical

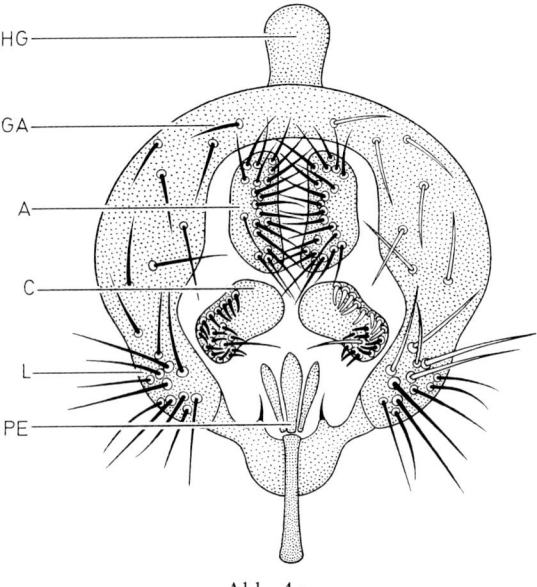

Abb. 4a

Fig. 4a—c. Cell lineage relationships in the male genital apparatus of *Drosophila*. a *in situ* (irradiated in second larval instar): a *yellow*-clone (white bristles) comprises part of a clasper *(C)*, lateral plates *(L)* and genital arch *(GA)*, but is confined to right side. b and c after duplication of a half disk: in b), "old" and "new" anal plates *(A)* are marked with a clone of *yellow* bristles. These are confined to the anal plates. In c, claspers *(C)* and lateral plates *(L)* of the "new" side share a common *yellow*-clone apparatus could be distinguished. A preceding analysis of the cell lineage relationships *in situ* had established that up to the early third instar the genitalia (claspers, lateral plates, and genital arch) share a common cell pool that appears already separated from the presumptive cells of the analia (anal plates) at the end of the first larval instar[2] (Fig. 4a). After symmetrical duplication, a clone on the "new" side was either con-

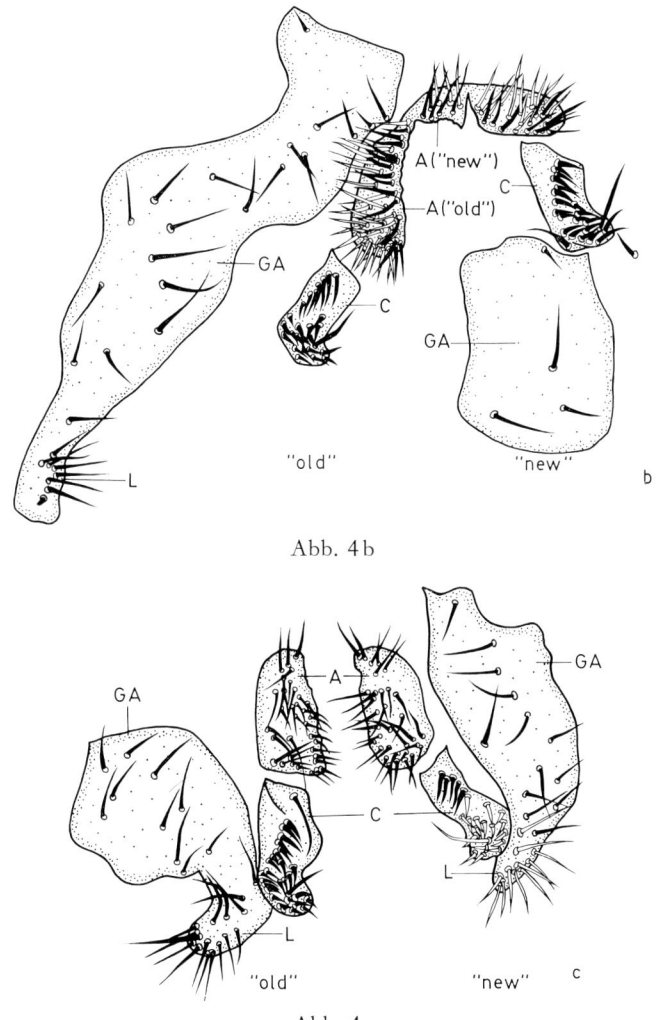

Abb. 4b

Abb. 4c

fined to the analia (Fig. 4b), or to the genitalia (Fig. 4c), but none of more than 20 clones comprised both groups. Among the genitalia, however, clones frequently embraced all three structures (*C, L, GA* in Fig. 4c). We have not encountered a

---

[2] The existence of these two separate cell populations is easily understood because analia and genitalia belong to different abdominal segments and in *Musca domestica* even form two separate imaginal disks (DÜBENDORFER, 1971).

situation where the analia were marked on the "old" side, and the genitalia on the "new" side, or *vice versa*. These results show that the blastema which forms the "new" side consists of two separate populations of cells, one determined to give rise to the analia, the other to the genitalia. Cells appear unable to change sides. However, at least three different pathways of development are still open to the presumptive cells of the genitalia. Parallel results were obtained with the female genital disk (ULRICH, 1971).

These observations may be interpreted as an indication for a prefinal state of determination of the presumptive cells of the male genitalia. The determination of claspers vs. lateral plates vs. genital arch must then only occur around the onset of metamorphosis. However, the same results may also be interpreted in SCHNEIDERMAN's terms (see POSTLETHWAIT, POODRY, and SCHNEIDERMAN, 1971; POODRY, BRYANT and SCHNEIDERMAN, 1971) to suggest that "repatterning" occurs, i.e. that cells assume new states of determination. According to this view, the clone mother cell may have been determined for clasper quality, e.g., or lateral plate quality, but changed its determination in the course of the cell divisions in the blastema[3].

We have to be cautious when we try to infer the state of determination of single cells *in situ* from the behavior of dissociated and regenerating imaginal disks. However, as long as our experiments fail to demonstrate that each cell in a disk is destined to develop into one particular bristle or trichome, these cells, by definition, are not yet definitively determined. Therefore, the results described above suggest to me that the cells in a mature disk are in a prefinal state of determination which in general is conservatively reproduced during a limited number of cell divisions (see GEHRING, this volume).

E. Determination Occurs in Groups of Cells

VOGT (1946) and recently GRIGLIATTI and SUZUKI (1971) have investigated temperature-sensitive alleles of the homoeotic mutation ss^a which causes parts of the antennal disk to develop into leg structures at 16° C, and into antennal structures at 25° C. The temperature-sensitive period lies in the third larval instar. In this stage the antennal disk is already clearly recognizable (see Fig. 12a, in VOGT, 1946), and must consist of a large number of cells (between 100 and 700, see Fig. 3). The change in the developmental pathway, viz. determination, which is brought about by a temperature shift, apparently takes place in a number of cells simultaneously. Large areas of the antenna are replaced by large sections of leg tissue.

A second example in which homoeotic mutations are not involved and which, therefore, has more bearing on determinative processes in normal development, has been described by BODENSTEIN and ABDEL-MALEK (1949). Wildtype larvae of *Drosophila virilis* were submerged for 30—45 min in a solution of nitrogen mustard. Up to 50% of the surviving adults responded with differentiation of leg structures at the expense of antennal structures if the treatment was given in the early third instar. Again, since at this time many cells are already present in the antennal disk, the transformation, i.e. the change in determination, most certainly involved a whole population of cells.

---

[3] A directed change of determination appears to occur regularly during true regeneration in the leg disk (SCHUBIGER, 1971; BRYANT, 1971).

In my opinion, the best evidence for our 5. statement can be derived from cell lineage studies. If mitotic recombination (see p. 5) is induced *after* a particular structure (e.g. the sex comb) or region (e.g. the basitarsus) has been determined, the marked clone will, of course, be confined to this structure/region. If it is induced prior to determination, the outcome will depend on whether the process of determination occurred in a single cell and was then clonally propagated *(a)*, or whether it occurred in a group of cells *(b)*. In *(a)*, a structure/region should consist completely either of marked, or of unmarked cells. In *(b)*, on the other hand, a structure/region may be mixed and contain both marked and unmarked cells. In both cases *(a)* and *(b)*, where the clone was induced prior to determination, it may also extend into adjacent differently determined structures/regions. When this simple logic is applied to published data (e.g. TOKUNAGA, 1962; BRYANT and SCHNEIDERMAN, 1969; BRYANT, 1970; POSTLETHWAIT and SCHNEIDERMAN, 1971a; GARCIA-BELLIDO and MERRIAM, 1971a), it becomes immediately clear that determination in normal development must affect cell groups. Fig. 5 shows a clone that runs through the basitarsus bisecting the

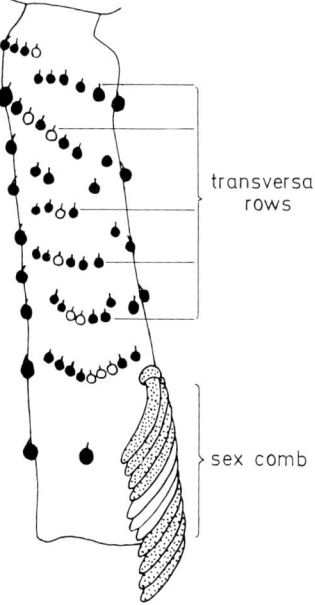

Fig. 5. Cell lineage relationships in the basitarsus of the male foreleg of *Drosophila*. A *yellow*-clone (white circles and bristles) extends longitudinally through all transversal rows and the sex comb. Circles represent the bristles of the transversal rows (After TOKUNAGA, 1962)

transversal rows and the sex comb. It so indicates that it was initiated before any of these structures became determined (see p. 4). Since the basitarsus, the transversal rows and the sex comb are mixed of marked and unmarked cells, they cannot themselves consist of one clone each. It follows that the determination for any of these structures occurred in more than one cell. The same conclusion can be reached for the determination of claspers and lateral plates in Fig. 4a.

A similar rationale has been used to show that the homoeotic determination in the mutant $Antp^R$ (POSTLETHWAIT and SCHNEIDERMAN, 1969, 1971b) and the process of transdetermination (GEHRING, this volume) most certainly are decisions that are reached by groups of cells.

It is an academic question whether determination is actually conveyed to a group of cells at once, or whether one cell becomes singularized and then communicates and transfers its state of determination to a number of selected neighbor cells. The experiment to distinguish between the two alternatives has yet to be devised.

F. The State of Determination in Disks of Mature Larvae is Very Stable Though Not Irreversible. It is Usually Propagated Conservatively to Daughter Cells

Experiments with dissociated disks from mature larvae of *Drosophila* have established that the cells, or small groups of cells, maintain their state of determination when isolated or trapped among differently determined cells (see p. 22; and GARCIA-BELLIDO, this volume). When fragments of mature disks are forced to undergo additional cell divisions by being cultured *in vivo* (see p. 4), this leads, as a rule, to a duplication (references in NÖTHIGER and SCHUBIGER, 1966), or multiplication (HADORN, 1963, 1966) of those primordia (viz. states of determination) that, according to the fate map, are present in the fragment (see p. 21, and Fig. 4). The stability of a given state of determination and its conservative mode of replication have been demonstrated in numerous investigations (see GEHRING, this volume).

We will now turn to experiments which reveal that under certain experimental conditions, disk primordia still possess an impressive developmental flexibility:

1) Various agents, such as nitrogen mustard (BODENSTEIN and ABDEL-MALEK, 1949), heavy doses of X-rays (WADDINGTON, 1942), or 5-fluoro-uracil (GEHRING, 1964), when applied to wildtype larvae are apparently capable of changing an established state of determination.

2) SCHUBIGER (1971) and BRYANT (1971) have recently provided convincing evidence that a particular region in the leg disk of late third instar larvae *(Drosophila)* has the capacity to regenerate the missing primordia so that a complete leg may be formed by the fragment. This phenomenon, in contrast to "simple" duplication or multiplication, requires the production of specific new states of determination. A discussion of the phenomenon of regeneration in imaginal disks may be found in papers by BRYANT (1971), and by GEHRING and NÖTHIGER (1972).

3) In long-term cultures of disk fragments *in vivo*, a given state of determination is usually reproduced during mitoses and passed on to the daughter cells. With a low probability, however, changes in the state of determination will occur. These "transdeterminations" will be discussed in a chapter of its own (GEHRING, this volume).

Acknowledgements

I want to thank my colleagues and friends, Drs. A. DÜBENDORFER, J. R. MERRIAM, H. TOBLER, and T. SPREIJ for their constructive criticism of the manuscript, and my wife for her patience.

References

ANDERSON, D. T.: The embryology of *Dacus tryoni*, 2. Development of imaginal discs in the embryo. J. Embryol. exp. Morph. **11**, 339—351 (1963).

ANDERSON, D. T.: The larval development of *Dacus tryoni* (Frogg.), 2. Development of imaginal rudiments other than the principle discs. Aust. J. Zool. **12**, 1—8 (1964).

Anderson, D. T.: The development of holometabolous insects. In: Waddington, C. H., Counce, S. J. (Eds.): Developmental Systems. Insects. New York-London: Academic Press (in press).

Auerbach, C.: The development of the legs, wings and halteres in wild type and some mutant strains of *Drosophila melanogaster*. Trans. roy. Soc. Edinb. **58**, 787–815 (1936).

Babcock, M. B.: Oviduct development in *Drosophila*. II. Metamorphic events in normal and ovariectomized females. Wilhelm Roux' Arch. Entwickl.-Meth. Org. **167**, 24–63 (1971).

Baker, W. K.: A clonal system of differential gene activity in *Drosophila*. Develop. Biol. **16**, 1–17 (1967).

Baker, W. K.: Position-effect variegation. Advanc. Genet. **14**, 133–169 (1968).

Bantock, C. R.: Experiments on chromosome elimination in the gall midge, *Mayetiola destructor*. J. Embryol. exp. Morph. **24**, 257–286 (1970).

Bautz, A.-M.: Chronologie de la mise en place de l'hypoderme imaginal de l'abdomen de *Calliphora erythrocephala* Meigen. Arch. Zool. exp. gén. **112**, 157–178 (1971).

Beadle, G. W., Tatum, E. L., Clancy, C. W.: Food level in relation to rate of development and eye pigmentation in *Drosophila melanogaster*. Biol. Bull. **75**, 447–462 (1938).

Becker, H. J.: On X-ray induced somatic crossing over. Drosophila Inform. Serv. **30**, 101 (1956).

Becker, H. J.: Über Röntgenmosaikflecken und Defektmutationen am Auge von *Drosophila* und die Entwicklungsphysiologie des Auges. Z. Vererb.-Lehre **88**, 333–373 (1957).

Becker, H. J.: The influence of heterochromatin, inversion heterozygosity and somatic pairing on X-ray induced mitotic recombination in *Drosophila melanogaster*. Molec. Gen. Genetics **105**, 203–218 (1969).

Bodenstein, D.: Investigations on the problem of metamorphosis. V. Some factors determining the facet number in the *Drosophila* mutant *Bar*. Genetics **24**, 494–508 (1939).

Bodenstein, D., Abdel-Malek, A.: The induction of *aristapedia* by nitrogen mustard in *Drosophila virilis*. J. exp. Zool. **111**, 95–115 (1949).

Bryant, P. J.: Cell lineage relationships in the imaginal wing disc of *Drosophila melanogaster*. Develop. Biol. **22**, 389–411 (1970).

Bryant, P. J.: Regeneration and duplication following operations *in situ* on the imaginal discs of *Drosophila melanogaster*. Develop. Biol. **26**, 606–615 (1971).

Bryant, P. J., Schneiderman, H. A.: Cell lineage, growth and determination in the imaginal leg discs of *Drosophila melanogaster*. Develop. Biol. **20**, 263–279 (1969).

Bull, A. L.: *Bicaudal*, a genetic factor which affects the polarity of the embryo in *Drosophila melanogaster*. J. exp. Zool. **161**, 221–242 (1966).

Cameron, I. L., Cleffmann, G.: Initiation of mitosis in relation to the cell cycle following feeding of starved chickens. J. Cell Biol. **21**, 169–174 (1964).

Chan, L. N., Gehring, W.: Determination of blastoderm cells in *Drosophila melanogaster*. Proc. nat. Acad. Sci. (Wash.) **68**, 2217–2221 (1971).

Chevais, S.: Déterminisme de la taille de l'oeil chez le mutant *Bar* de la Drosophile. Intervention d'une substance diffusible spécifique. Bull. biol. france belg. **78**, 71–110 (1944).

Clark, A. M., Gould, A. B., Graham, S. F.: Patterns of development among mosaics in *Habrobracon juglandis*. Develop. Biol. **25**, 133–148 (1971).

Counce, S. J.: The causal analysis of insect embryogenesis. In: Waddington, C. H., Counce, S. J. (Eds.): Developmental Systems. Insects. New York-London: Academic press (in press).

Dahlhelm, D.: Die Entwicklung des Integumentes bei der Larve von *Calliphora erythrocephala* Meigen. Biol. Zbl. **86**, 141–175 (1967).

Dübendorfer, A.: Untersuchungen zum Anlageplan und Determinationszustand der weiblichen Genital- und Analprimordien von *Musca domestica* L. Wilhelm Roux' Arch. Entwickl.-Mech. Org. **168**, 142–168 (1971).

El Shatoury, H. H.: The structure of the lymph glands of *Drosophila* larvae. Wilhelm Roux' Arch. Entwickl. Mech. Org. **147**, 489–495 (1955).

Ephrussi, B., Beadle, G. W.: A technique of transplantation for *Drosophila*. Amer. Naturalist **70**, 218–225 (1936).

Fischer, J., Rosin, S.: Das larvale Wachstum von *Chironomus nuditarsis* Str. Rev. Suisse Zool. **76**, 727–734 (1969).

Fristrom, D.: Cellular degeneration in the production of some mutant phenotypes in *Drosophila melanogaster*. Molec. Gen. Genetics **103**, 363—379 (1969).

Fristrom, D.: Chemical modification of cell death in the *Bar* eye of *Drosophila*. Molec. Gen. Genetics **115**, 10—18 (1972).

Fristrom, J.W.: The biochemistry of imaginal disk development. This volume.

Garcia-Bellido, A.: Pattern reconstruction by dissociated imaginal disc cells of *Drosophila melanogaster*. Develop. Biol. **14**, 278—306 (1966).

Garcia-Bellido, A.: Histotypic reaggregation of dissociated imaginal disc cells of *Drosophila melanogaster* cultured *in vivo*. Wilhelm Roux' Arch. Entwickl.-Mech. Org. **158**, 212—217 (1967).

Garcia-Bellido, A.: Some parameters of mitotic recombination in *Drosophila melanogaster*. Molec. Gen. Genetics **115**, 54—72 (1972).

Garcia-Bellido, A.: Pattern formation in imaginal disks. This volume.

Garcia-Bellido, A., Merriam, J.R.: Cell lineage of the imaginal discs in *Drosophila* gynandromorphs. J. exp. Zool. **170**, 61—76 (1969a).

Garcia-Bellido, A., Merriam, J.R.: A preliminary morphogenetic map of the wing disc. Drosophila Inform. Serv. **44**, 65 (1969b).

Garcia-Bellido, A., Merriam, J.R.: Parameters of the wing imaginal disc development of *Drosophila melanogaster*. Develop. Biol. **24**, 61—87 (1971a).

Garcia-Bellido, A., Merriam, J.R.: Clonal parameters of tergite development in *Drosophila*. Develop. Biol. **26**, 264—276 (1971b).

Gateff, E.: Developmental capacities of immature eye-antennal discs of *Drosophila*. Ph.D. thesis, University of California, Irvine, USA (1971).

Gateff, E., Schneiderman, H.A.: Neoplasms in mutant and cultured wild-type tissues of *Drosophila*. Nat. Cancer Inst. Monogr. **31**, 365—397 (1969).

Gehring, W.: Phenocopies produced by 5-fluoro-uracil. Drosophila Inform. Serv. **39**, 102 (1964).

Gehring, W.: The stability of the determined state in cultures of imaginal disks in *Drosophila*. This volume.

Gehring, W., Nöthiger, R.: The imaginal discs of *Drosophila*. In: Waddington, C.H., Counce, S.J. (Eds.): Developmental Systems. Insects. New York-London: Academic Press (in press).

Geigy, R.: Erzeugung rein imaginaler Defekte durch ultraviolette Eibestrahlung bei *Drosophila melanogaster*. Wilhelm Roux' Arch. Entwickl.-Mech. Org. **125**, 406—447 (1931a).

Geigy, R.: Action de l'ultraviolet sur le pole germinal dans l'oeuf de *Drosophila melanogaster* (castration et mutabilité). Rev. Suisse Zool. **38**, 187—288 (1931b).

Geyer-Duszynska, I.: Experimental research on chromosome elimination in *Cecidomyidae* (*Diptera*). J. exp. Zool. **141**, 391—447 (1959).

Gilbert, L.I., Schneiderman, H.A.: Control of growth and development in insects. Science **143**, 325—333 (1964).

Gloor, H.: Phänokopie-Versuche mit Aether an *Drosophila*. Rev. Suisse Zool. **54**, 637—712 (1947).

Grigliatti, T., Suzuki, D.T.: Temperature-sensitive mutations in *Drosophila melanogaster*, VIII. The homeotic mutant, ss^{a40a}. Proc. nat. Acad. Sci. (Wash.) **68**, 1307—1311 (1971).

Gsell, R.: Untersuchungen zur Stabilität einer *yellow* Positionseffekt-Variegation in Imaginalscheiben-Kulturen von *Drosophila melanogaster*. Molec. Gen. Genetics **110**, 218—237 (1971).

Gustafson, T., Wolpert, L.: Cellular movement and contact in sea urchin morphogenesis. Biol. Rev. **42**, 442—498 (1967).

Hadorn, E.: Regulation and differentiation within field districts in imaginal discs of *Drosophila*. J. Embryol. exp. Morph. **1**, 213—216 (1953).

Hadorn, E.: Differenzierungsleistungen wiederholt fragmentierter Teilstücke männlicher Genitalscheiben von *Drosophila melanogaster* nach Kultur *in vivo*. Develop. Biol. **7**, 617—629 (1963).

Hadorn, E.: Problems of determination and transdetermination. In "Genetic control of differentiation". Brookhaven Symp. Biol. **18**, 148—161 (1965).

Hadorn, E.: Konstanz, Wechsel und Typus der Determination und Differenzierung in Zellen aus männlichen Genitalanlagen von *Drosophila melanogaster* nach Dauerkultur *in vivo*. Develop. Biol. **13**, 424—509 (1966).

Hadorn, E., Anders, G., Ursprung, H.: Kombinate aus teilweise dissoziierten Imaginalscheiben verschiedener Mutanten und Arten von *Drosophila*. J. exp. Zool. **142**, 159—175 (1959).

Hadorn, E., Bertani, G., Gallera, J.: Regulationsfähigkeit und Feldorganisation der männlichen Genital-Imaginalscheibe von *Drosophila melanogaster*. Wilhelm Roux' Arch. Entwickl.-Mech. Org. **144**, 31—70 (1949).

Hadorn, E., Garcia-Bellido, A.: Zur Proliferation von *Drosophila* Zellkulturen im Adultmilieu. Rev. Suisse Zool. **71**, 576—582 (1964).

Hadorn, E., Gsell, R., Schultz, J.: Stability of a position-effect variegation in normal and transdetermined larval blastemas from *Drosophila melanogaster*. Proc. Nat. Acad. Sci., (Wash.) **65**, 633—637 (1970).

Hadorn, E., Hürlimann, R., Mindek, G., Schubiger, G., Staub, M.: Entwicklungsleistungen embryonaler Blasteme von *Drosophila* nach Kultur im Adultwirt. Rev. Suisse Zool. **75**, 557—569 (1968).

Haendle, J.: Röntgeninduzierte mitotische Rekombination bei *Drosophila melanogaster*. I. Ihre Abhängigkeit von der Dosis, der Dosisrate und vom Spektrum. Molec. Gen. Genetics **113**, 114—131 (1971 a).

Haendle, J.: Röntgeninduzierte mitotische Rekombination bei *Drosophila melanogaster*. II. Beweis der Existenz und Charakterisierung zweier von der Art des Spektrums abhängiger Reaktionen. Molec. Gen. Genetics **113**, 132—149 (1971 b).

Henke, K., Maas, H.: Über sensible Perioden der allgemeinen Körpergliederung von *Drosophila*. Nachr. Akad. Wiss. Göttingen, Math.-phys. Kl. **1**, 3—4 (1946).

Hinton, C. W.: The behaviour of an unstable ring-chromosome of *Drosophila melanogaster*. Genetics **40**, 951—961 (1955).

Howland, R. B., Child, G. P.: Experimental studies on development in *Drosophila melanogaster*. I. Removal of protoplasmic materials during late cleavage and early embryonic stages. J. exp. Zool. **70**, 415—427 (1935).

Howland, R. B., Sonnenblick, B. P.: Experimental studies on development in *Drosophila melanogaster*. II. Regulation in the early egg. J. exp. Zool. **73**, 109—125 (1936).

Illmensee, K.: Transplantation of embryonic nuclei into unfertilized eggs of *Drosophila melanogaster*. Nature (Lond.) **219**, 1268—1269 (1968).

Illmensee, K.: Imaginal structures after nuclear transplantation in *Drosophila melanogaster*. Naturwissenschaften **11**, 550—551 (1970).

Illmensee, K.: Developmental potencies of nuclei from cleavage, preblastoderm, and syncytial blastoderm transplanted into unfertilized eggs of *Drosophila melanogaster*. Wilhelm Roux' Arch. Entwickl.-Mech. Org. (in press).

Janning, W.: Bestimmung des Heterochromatisierungsstadiums beim *white*-Positionseffekt mittels röntgeninduzierter mitotischer Rekombination in der Augenanlage von *Drosophila melanogaster*. Molec. Gen. Genetics **107**, 128—149 (1970).

Kalthoff, K.: Position of targets and period of competence for UV-induction of the malformation "double abdomen" in the egg of *Smittia* spec. (*Diptera, Chironomidae*). Wilhelm Roux' Arch. Entwickl.-Mech. Org. **168**, 63—84 (1971 a).

Kalthoff, K.: Photoreversion of UV-induction of the malformation "double abdomen" in the egg of *Smittia* spec. (*Diptera, Chironomidae*). Develop. Biol. **25**, 119—132 (1971 b).

Karfunkel, P.: The role of microtubules and microfilaments in neurulation in *Xenopus*. Develop. Biol. **25**, 30—56 (1971).

Laugé, G.: Recherches expérimentales sur la détermination et la différenciation des caractères morphologiques et histologiques des intersexués triploides de *Drosophila melanogaster* Meig. Ann. Embryol. Morph. **2**, 245—270 (a), and 273—299 (b) (1969).

Lawrence, P. A.: Development and determination of hairs and bristles in the milkweed bug, *Oncopeltus fasciatus* (*Lygaeidae, Hemiptera*). J. Cell Sci. **1**, 475—498 (1966).

Lees, A. D., Waddington, C. H.: The development of the bristles in normal and some mutant types of *Drosophila melanogaster*. Proc. Roy. Soc. B (Edinb.) **131**, 87—110 (1942).

Lindsley, D. L., Grell, E. H.: Genetic variations of *Drosophila melanogaster*. Carnegie Inst. Wash. Publ. **627** (1968).

Löbbecke, E. A.: Autoradiographische Bestimmung der DNS-Synthese-Dauer von Zellen der Flügelimaginalanlage von *Ephestia kühniella*. Z. Wilhelm Roux' Arch. Entwickl.-Mech. Org. **162**, 1—18 (1969).

Lüönd, H.: Untersuchungen zur Mustergliederung in fragmentierten Primordien des männlichen Geschlechtsapparates von *Drosophila séguyi*. Develop. Biol. **3**, 615—656 (1961).

Mahowald, A. P.: Polar granules of *Drosophila*. II. Ultrastructural changes during early embryogenesis. J. exp. Zool. **167**, 237—261 (1968).

Mahowald, A. P.: Origin and continuity of polar granules. In: Results and problems in cell differentiation, Vol. 2, pp. 158—169, Berlin-Heidelberg-New York: Springer 1971.

Mandaron, P.: Développement *in vitro* des disques imaginaux de la Drosophile. Aspects morphologiques et histologiques. Develop. Biol. **22**, 298—320 (1970).

Mandaron, P.: Sur le mécanisme de l'évagination des disques imaginaux chez la Drosophile. Develop. Biol. **25**, 581—605 (1971).

Merriam, J. R., Fyffe, W. E.: Somatic crossing over in *Drosophila melanogaster*: I. Dose response curves for X-ray induction and effects of dose fractionation. Mutation Res. **14**, 309—314 (1972).

Merriam, J. R., Garcia-Bellido, A.: A model for somatic pairing derived from somatic crossing over with third chromosome rearrangements in *Drosophila melanogaster*. Molec. Gen. Genetics **115**, 302—313 (1972).

Merriam, J. R., Nöthiger, R., Garcia-Bellido, A.: Are dicentric anaphase bridges formed by somatic recombination in X chromosome inversion heterozygotes of *Drosophila melanogaster*? Molec. Gen. Genetics **115**, 294—301 (1972).

Mindek, G.: Metamorphosis of imaginal discs of *Drosophila melanogaster*. Wilhelm Roux' Arch. Entwickl.-Mech. Org. **169**, 353—356 (1972).

Mindek, G., Nöthiger, R.: Parameters influencing the acquisition of competence for metamorphosis in imaginal disks of *Drosophila* (In prep.).

Murphy, C.: Determination of the dorsal mesothoracic disc in *Drosophila*. Develop. Biol. **15**, 368—394 (1967).

Muth, F. W.: Untersuchungen zur Wirkungsweise der Mutante "*kfl*" bei der Mehlmotte *Ephestia kühniella* Z. Wilhelm Roux' Arch. Entwickl.-Mech. Org. **153**, 370—418 (1961).

Nesbitt, M. N.: X chromosome inactivation mosaicism in the mouse. Develop. Biol. **26**, 252—263 (1971).

Newby, W. W.: A study of intersexes produced by a dominant mutation in *Drosophila virilis*, Blanco Stock. Univ. Texas Publ. **4228**, 113—145 (1942).

Nöthiger, R.: Differenzierungsleistungen in Kombinaten, hergestellt aus Imaginalscheiben verschiedener Arten, Geschlechter und Körpersegmente von *Drosophila*. Wilhelm Roux' Arch. Entwickl.-Mech. Org. **155**, 269—301 (1964).

Nöthiger, R., Dübendorfer, A.: Somatic crossing-over in the housefly. Molec. Gen. Genetics **112**, 9—13 (1971).

Nöthiger, R., Oberlander, H.: Differentiation of pulsating regions in genital imaginal discs cultured *in vivo* (*Drosophila melanogaster*). J. exp. Zool. **164**, 61—68 (1967).

Nöthiger, R., Schubiger, G.: Developmental behaviour of fragments of symmetrical and asymmetrical imaginal discs of *Drosophila melanogaster* (*Diptera*). J. Embryol. exp. Morph. **16**, 355—368 (1966).

Nöthiger, R., Strub, S.: Imaginal defects after UV-microbeam irradiation of early cleavage stages of *Drosophila melanogaster*. Rev. Suisse Zool. **79**, 267—279 (1972).

Nöthiger, R., Ulrich, E.: Cell lineage and determination in the male genital disk of *Drosophila melanogaster*. (In prep.).

Oberlander, H.: Hormonal control of disk development. This volume.

Oelhafen, F.: Zur Embryogenese von *Culex pipens*: Markierungen und Exstirpationen mit UV-Strahlenstich. Wilhelm Roux' Arch. Entwickl.-Mech. Org. **153**, 120—157 (1961).

Okada, E., Waddington, C. H.: The submicroscopic structure of the *Drosophila* egg. J. Embryol. exp. Morph. **7**, 583—597 (1959).

Patterson, J. T., Stone, W.: Gynandromorphs in *Drosophila melanogaster*. Univ. Texas Publ. **3825**, 1—67 (1938).

Pentz, S., Krause, G.: Nester aberranter Schuppen nach R-Bestrahlung der Flügelanlagen in weiblichen Diapause- und Nondiapause-Raupen von *Plodia* (*Lepidoptera*). Wilhelm Roux' Arch. Entwickl.-Mech. Org. **160**, 167—186 (1968).

Perry, M. M.: Further studies on the development of the eye of *Drosophila melanogaster*. I. The ommatidia. J. Morph. **124**, 227—148 (1968).

Poodry, C. A., Bryant, P., Schneiderman, H. A.: The mechanism of pattern reconstruction by dissociated imaginal discs of *Drosophila melanogaster*. Develop. Biol. **26**, 464—477 (1971).

Poodry, C. A., Schneiderman, H. A.: The ultrastructure of the developing leg of *Drosophila melanogaster*. Wilhelm Roux' Arch. Entwickl.-Mech. Org. **166**, 1—44 (1970).

Poodry, C. A., Schneiderman, H. A.: Intercellular adhesivity and pupal morphogenesis in *Drosophila melanogaster*. Wilhelm Roux' Arch. Entwickl.-Mech. Org. **168**, 1—9 (1971).

Postlethwait, J. H., Poodry, C. A., Schneiderman, H. A.: Cellular dynamics of pattern duplication in imaginal discs of *Drosophila melanogaster*. Develop. Biol. **26**, 125—132 (1971).

Postlethwait, J. H., Schneiderman, H. A.: A clonal analysis of determination in *Antennapedia*, a homoeotic mutant of *Drosophila melanogaster*. Proc. nat. Acad. Sci. (Wash.) **64**, 176—183 (1969).

Postlethwait, J. H., Schneiderman, H. A.: A clonal analysis of development in *Drosophila melanogaster*: morphogenesis, determination and growth in the wild-type antenna. Develop. Biol. **24**, 477—519 (1971a).

Postlethwait, J. H., Schneiderman, H. A.: Pattern formation and determination in the antenna of the homoeotic mutant *Antennapedia* of *Drosophila melanogaster*. Develop. Biol. **25**, 606—640 (1971b).

Ripoll, P.: The embryonic organization of the imaginal wing disc of *Drosophila melanogaster*. Wilhelm Roux' Arch. Entwickl.-Mech. Org. **169**, 200—215 (1972).

Saunders, J. W. Jr., Fallon, J. F.: Cell death in morphogenesis. In: Locke, M. (Ed.): Major Problems in Developmental Biology, p. 289—314. New York-London: Academic Press 1966.

Schläpfer, T.: Der Einfluß des adulten Wirtsmilieus auf die Entwicklung von larvalen Augenantennen-Imaginalscheiben von *Drosophila melanogaster*. Wilhelm Roux' Arch. Entwickl.-Mech. Org. **154**, 378—404 (1963).

Schneider, I.: Cell lines derived from late embryonic stages of *Drosophila melanogaster*. J. Embryol. exp. Morph. **27**, 353—365 (1972).

Schubiger, G.: Anlageplan, Determinationszustand und Transdeterminationsleistungen der männlichen Vorderbeinscheibe von *Drosophila melanogaster*. Wilhelm Roux' Arch. Entwickl.-Mech. Org. **160**, 9—40 (1968).

Schubiger, G.: Regeneration, duplication and transdetermination in fragments of the leg disc of *Drosophila melanogaster*. Develop. Biol. **26**, 277—295 (1971).

Schubiger, M., Schneiderman, H. A.: Nuclear transplantation in *Drosophila melanogaster*. Nature (Lond.) **230**, 185—186 (1971).

Schweizer, P.: Wirkung von Röntgenstrahlen auf die Entwicklung der männlichen Genitalprimordien von *Drosophila melanogaster* und Untersuchung von Erholungsvorgängen durch Zellklon-Analyse. Biophysik **8**, 158—188 (1972).

Shaaya, E., Karlson, P.: Der Ecdysontiter während der Insektenentwicklung. II. Die postembryonale Entwicklung der Schmeißfliege *Calliphora erythrocephala* Meig. J. Insect. Physiol. **11**, 65—69 (1965).

Shearn, A., Rice, T., Garen, A., Gehring, W.: Imaginal disc abnormalities in lethal mutants of *Drosophila*. Proc. nat. Acad. Sci. (Wash.) **68**, 2594—2598 (1971).

Spreij, T. E.: Cell death during the development of the imaginal disks of *Calliphora erythrocephala*. Neth. J. Zool. **21**, 221—264 (1971).

Stern, C.: Somatic crossing over and segregation in *Drosophila melanogaster*. Genetics **21**, 625—730 (1936).

Stern, C.: The prospective significance of imaginal discs in *Drosophila*. J. Morph. **67**, 107—122 (1940).

Stern, C.: Genetic Mosaics and Other Essays. Cambridge: Harvard University Press 1968.

Stewart, R. N., Dermen, H.: Determination of number and mitotic activity of shoot apical initial cells by analysis of mericlinal chimeras. Amer. J. Bot. **57**, 816—826 (1970).

Stewart, M., Murphy, C., Fristrom, J. W.: The recovery and preliminary characterization of X chromosome mutants affecting imaginal discs of *Drosophila melanogaster*. Develop. Biol. **27**, 71—83 (1972).

Stumpf, H.: Die Richtungen der Teilungsspindeln auf dem Puppenflügel von *Drosophila* im Verlaufe der Mitosenperiode. Biol. Zbl. **75**, 17—27 (1956).

Sturtevant, A. H.: The *claret* mutant type of *Drosophila simulans*: a study of chromosome elimination and of cell-lineage. Z. wiss. Zool. **135**, 324—355 (1929).

Tettenborn, U., Dofuku, R., Ohno, S.: Noninducible phenotype exhibited by a proportion of female mice heterozygous for the X-linked testicular feminisation mutation. Nature New Biol. **234**, 37—40 (1971).

Timm, U.: Quantitative Untersuchungen über das Zellteilungswachstum der Flügelanlagen von *Ephestia kühniella* Zeller in Regenerationsexperimenten. Thesis, Math.-naturw. Fakultät der Universität Köln, Köln, W-Germany (1970).

Tobler, H.: Zellspezifische Determination und Beziehung zwischen Proliferation und Transdetermination in Bein- und Flügelprimordien von *Drosophila melanogaster*. J. Embryol. exp. Morph. **16**, 609—633 (1966).

Tokunaga, C.: Cell lineage and differentiation on the male foreleg of *Drosophila melanogaster*. Develop. Biol. **4**, 489—516 (1962).

Trinkaus, J. P.: Morphogenetic cell movements. In: Locke, M. (Ed.): Major Problems in Developmental Biology, pp. 125—176. New York-London: Academic Press 1966.

Ulrich, E.: Cell lineage, Determination und Regulation in der weiblichen Genitalimaginalscheibe von *Drosophila melanogaster*. Wilhelm Roux' Arch. Entwickl.-Mech. Org. **167**, 64—82 (1971).

Ursprung, H.: Fragmentierungs- und Bestrahlungsversuche zur Bestimmung von Determinationszustand und Anlageplan der Genitalscheiben von *Drosophila melanogaster*. Wilhelm Roux' Arch. Entwickl.-Mech. Org. **151**, 504—558 (1959).

Ursprung, H.: Der Einfluß des Wirtsalters auf die Entwicklungsleistung von Sagittalhälften männlicher Genitalscheiben von *Drosophila melanogaster*. Develop. Biol. **4**, 22—39 (1962).

Ursprung, H.: The fine structure of imaginal disks. This volume.

Ursprung, H., Schabtach, E.: The fine structure of the male *Drosophila* genital disk during late larval and early pupal development. Wilhelm Roux' Arch. Entwickl.-Mech. Org. **160**, 243—254 (1968).

Vig, B. K., Paddock, E. F.: Studies on the expression of somatic crossing over in *Glycine max*. L. Theoret. Appl. Genetics **40**, 316—321 (1970).

Vogt, M.: Zur labilen Determination der Imaginalscheiben von *Drosophila*. II. Die Umwandlung präsumptiven Fühlergewebes in Beingewebe. Biol. Zbl. **65**, 238—254 (1946).

Waddington, C. H.: Some developmental effects of X-rays in *Drosophila*. J. exp. Biol. **19**, 101—117 (1942).

Waddington, C. H.: New Patterns in Genetics and Development, p. 213. New York-London: Columbia University Press 1962.

Waddington, C. H.: Fields and gradients. In: Locke, M. (Ed.): Major Problems in Developmental Biology, pp. 105—124. New York-London: Academic Press 1966.

Waddington, C. H., Perry, M. M.: The ultrastructure of the developing eye of *Drosophila*. Proc. roy. Soc. B. (Edinb.) **153**, 155—178 (1960).

Wald, H.: Cytological studies on abnormal development of eggs of the *claret* mutant type of *Drosophila simulans*. Genetics **21**, 264—281 (1936).

Walen, K. H.: Somatic crossing over in relationship to heterochromatin in *Drosophila melanogaster*. Genetics **49**, 905—923 (1964).

Wehman, H. J.: Fine structure of *Drosophila* wing imaginal discs during early stages of metamorphosis. Wilhelm Roux' Arch. Entwickl.-Mech. Org. **163**, 375—390 (1969).

Wildermuth, H. R.: Differenzierungsleistungen, Mustergliederung und Transdeterminationsmechanismen in hetero- und homoplastischen Transplantaten der Rüsselprimordien von *Drosophila*. Wilhelm Roux' Arch. Entwickl.-Mech. Org. **160**, 41—75 (1968).

Wildermuth, H. R.: Determination and transdetermination in cells of the fruitfly. Sci. Prog. Oxf. **58**, 329—358 (1970).

Wilt, F. H., Wessells, N. K.: Methods in Developmental Biology. New York: T. Y. Crowell Comp. 1967.

Zalokar, M.: L'ablation des disques imaginaux chez la larve de Drosophile. Rev. Suisse Zool. **50**, 232—236 (1943).

Zalokar, M.: Transplantation of nuclei in *Drosophila melanogaster*. Proc. nat. Acad. Sci. (Wash.) **68**, 1539—1541 (1971).

The Stability of the Determined State in Cultures of Imaginal Disks in Drosophila[1]

WALTER GEHRING

Department of Anatomy, Yale University, New Haven, Connecticut

I. Determination and Differentiation

The development of different cell types in higher organisms, arising from one cell type (the zygote), is designated as *cell differentiation*. Differentiation is initiated by the process of *determination* which programs the cells for their future developmental pathway. There is accumulating evidence that the developmental programs reside in the genome and represent sets of genes acting in a coordinate fashion. The fundamental problem of how different genes are coordinately controlled remains to be elucidated. Two major categories of determinative processes are known: one is *embryonic induction* which involves cell or tissue interactions by means of diffusible inducing substances, and another is *cytoplasmic segregation* which is based upon factors intrinsic to the differentiating cells. In the latter case differentiation may be achieved by unequal distribution during cell division of cell components acting as determinative factors. Since determination creates the first essential differences among cells, an understanding of this process might be a key to the problem of differentiation in general.

The *imaginal disks* of *Drosophila* are a highly suitable material for the study of determination, since a method has been devised for culturing disk cells permanently in a determined state (HADORN, 1963). In addition, *Drosophila* offers the possibility of an embryological and biochemical as well as a genetic investigation. Hitherto, the emphasis was mainly on the embryological and genetic approach which will be discussed in this survey.

The aim of this chapter is to give a brief review of the experimental results so far obtained and to discuss the conclusions that can be drawn from them. The latter may still be preliminary but they will perhaps stimulate further investigations in the field. The main topic concerns the stability of the determined state and the qualitative changes of determination occurring in cultured cells, since these phenomena might give the insight into the problem of determination. Hopefully some of the readers might become "determined" to enter this field, which offers ample possibilities for an experimental approach to this intriguing problem.

---

[1] This is an updated and revised version of an article which appeared in Vol. 1 of this series. The more recent experiments reported in this article were supported by a grant from the Jane Coffin Childs Memorial Fund for Medical Research, and by the National Science Foundation (Grant GB-17267 X 2).

II. Imaginal Disks and their Culture in vivo

The *Drosophila* larva contains in addition to its larval structures a number of primordia, called *imaginal disks*, which during metamorphosis give rise to specific structures of the adult fly. At the pupal stage most of the larval tissues are histolyzed and the imago (adult) is formed anew from the imaginal primordia (see BODENSTEIN, 1950; GEHRING and NÖTHIGER, in press). There are three pairs of disks forming the

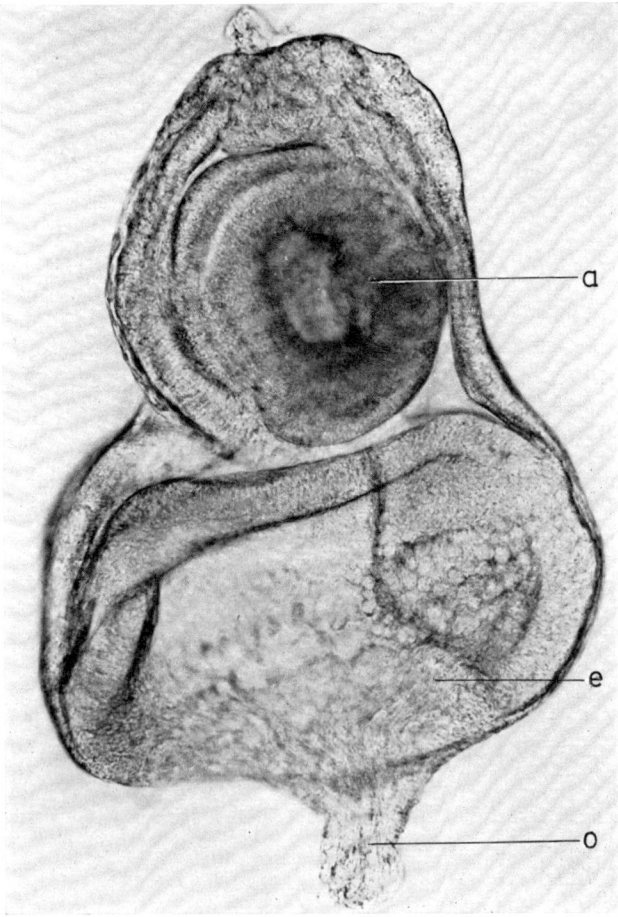

Fig. 1. Eye-antennal disk of a mature larva. Living unstained preparation. *a* antennal disk, *e* eye disk, *o* optic stalk. (Magnification 200×)

head of the fly: the labial disks, the imaginal cells of the clypeo-labrum (GEHRING and SEIPPEL, 1967) and the eye-antennal disks (Fig. 1). The adult *thorax* and its appendages arise from three pairs of leg disks and three pairs of dorsal thoracic disks forming the dorsal part of the thorax with the wings and the halteres. Finally, the integument of the *abdomen* is derived from small groups of imaginal cells called "histoblasts", and the genital apparatus located in the last abdominal segments is formed by the unpaired genital disk. Separate primordia have been identified for several internal organs like the gut, the salivary glands and the gonads.

The imaginal disks first appear as invaginations or thickenings of the epidermis in the later embryonic stages. It is not known whether all the disk cells are of ectodermal origin or whether there is a contribution from invading mesodermal cells as well. During the larval stages the disks grow considerably by cell division, acquiring a definite shape characteristic for each kind of disk; differentiation into adult structures is delayed until pupation. The disks consist of a single epithelial cell layer which becomes folded in the course of the larval development and encloses a narrow lumen. The outer surface of the disk is covered by a basement lamina. The disk cells are small basophilic cells about 5 μ in diameter, with a relatively large nucleus and little surrounding cytoplasm. In Fig. 2 living cells of a dissociated eye-antennal disk are shown in phase contrast. The cells appear fairly uniform, even at the ultrastructural

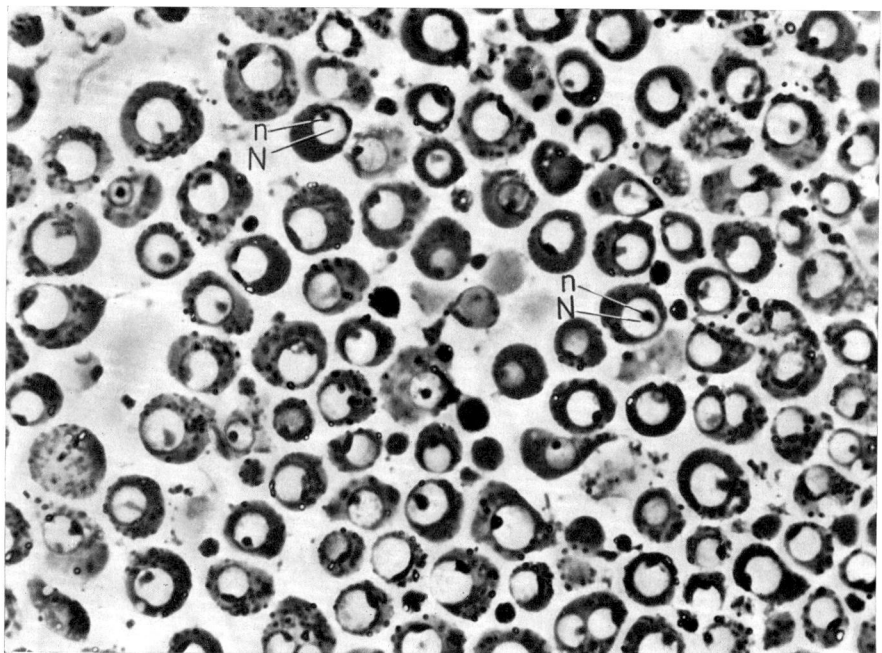

Fig. 2. Living cells from a dissociated eye-antennal disk. Phase contrast. N nucleus, n nucleolus. (Magnification 800×). By courtesy of P. SUTTER, P. FELS, and H. FRISCHKNECHT

level (see URSPRUNG, this volume). WADDINGTON and PERRY (1960) have described clusters of cells in the eye disk which appear to represent the anlagen of the ommatidia and can be distinguished cytologically from the remaining disk cells. Otherwise, there is little if any morphological differentiation among the disk cells.

Disks can be transplanted into larval or adult hosts by a method devised by EPHRUSSI and BEADLE (1936) (see also URSPRUNG, 1967). The donor larva is dissected and the isolated disk is injected by means of a micropipette into the body cavity of a host animal. When a disk is transplanted into a *larval host* of the same age as the donor, it develops synchronously with the host and undergoes metamorphosis. After the host fly has hatched, the implant can be removed from its abdomen and the structures formed by the implant examined.

Using the method of HADORN (1963) blastemas of imaginal disks can be cultured indefinitely in the abdomen of adult flies where they do not undergo metamorphosis. The host's hemolymph serves as a culture medium which allows proliferation, but

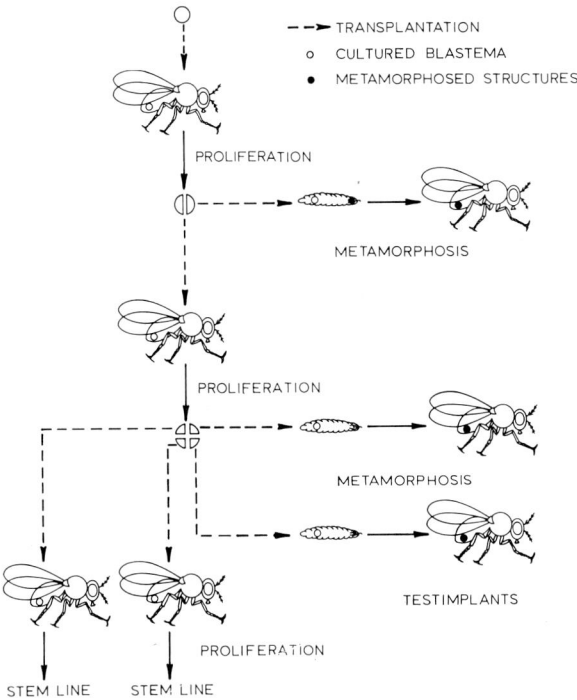

Fig. 3. Method for culturing imaginal disks *in vivo* (modified after HADORN, 1963). Explanation in the text

due to a low level or absence of the moulting hormone adult differentiation is not induced. Thus, the implant maintains its larval state. Since the culture period is limited by the lifespan of the host fly, the implant must be transferred to a young adult host every 2 to 4 weeks (Fig. 3). At each transfer the implant, which has increased in size, is cut into a number of fragments. Some of these fragments are injected into fresh adult hosts (stem line), whereas others are transplanted back into larvae with which they undergo metamorphosis. These larval "test implants" provide information about the capacities of the cultured disk cells for differentiation. It has been found that the cultured cells maintain their capacity for normal differentiation even after several years of culturing (HADORN, 1966) and also show the normal karyotype (GEHRING, 1966a; REMENSBERGER, 1968). Occasionally, loss of the capacity to differentiate and abnormal karyotypes are observed. It should be emphasized that we are dealing with cultures of blastemas rather than cultures of separate cells. However, there are techniques available to study the capacities of single cells in the cultures as well. Isolated cells obtained by dissociation of disks carrying different marker genes can be intermixed, reaggregated, and tested for their developmental behavior (HADORN, ANDERS, and URSPRUNG, 1959; NÖTHIGER, 1964; GARCIA-BELLIDO, 1966a). Another method makes use of somatic crossing-over or somatic mutation to induce genetically marked single cells which will form clones during proliferation in culture (GEHRING, 1967). Thus, it is possible to analyze determination at the cellular level, using *in vivo* techniques.

III. Determination of Imaginal Disks and their Progenitor Cells

The following operational criteria have been used to demonstrate the state of determination of imaginal disc cells experimentally: 1) their capacity to differentiate autonomously when isolated and intermixed with cells of a different type; 2) their association in cell mixtures with cells of the same histotype and segregation from cells of a different histotype; and 3) their capacity to transmit a specific state of determination clonally to their descendants. The clonal differentiation has been studied extensively in genetic mosaics. Genetically marked nuclei or cells are produced by somatic mutation (STURTEVANT, 1929) or somatic crossing-over (STERN, 1936). In the course of proliferation the marked cell gives rise to a clone that can be detected in the adult as a patch of marked tissue. Gynandromorphs produced by chromosome elimination in one of the first cleavage divisions provide evidence for the totipotency of early cleavage nuclei (STURTEVANT, 1929; GARCIA-BELLIDO and MERRIAM, 1969). The borderline between XX (female) and X0 (male) cells in adult gynandromorphes is more or less random indicating that the early cleavage nuclei can give rise to any part of the adult. The potential of later cleavage nuclei has been tested by nuclear transplantation (GEYER-DUSZYNSKA, 1967; ILLMENSEE, 1968 and 1970; SCHUBIGER and SCHNEIDERMAN, 1971; ZALOKAR, 1971). ILLMENSEE (pers. comm.) succeeded in injecting genetically marked single nuclei with some adhering cytoplasm from syncytial stages into unfertilized eggs. A significant fraction of the eggs developed into fairly normal larvae of the donor genotype; and there was no difference in the potential of nuclei taken from various regions of the syncytial embryo. Furthermore, ZALOKAR (1971) obtained mosaic flies by injecting genetically marked cleavage nuclei with adhering cytoplasm into cleavage stage embryos. Among the mosaic animals which developed to the adult stage, two males were found which produced fertile sperm carrying the donor genotype. Since the donor nuclei were not taken from the posterior pole of the cleavage embryo where the primordial germ cells are formed, this observation lends further support to the idea that the cleavage nuclei when tested under these conditions are totipotent.

At the blastoderm stage, when cell membranes are first formed, the cells are already restricted in their potential for forming adult epidermal structures (CHAN and GEHRING, 1971). Based upon the work of HADORN and collaborators (HADORN et al., 1968; SCHUBIGER, SCHUBIGER, and HADORN, 1969) on cultures of embryonic tissue, a method for culturing blastoderm cells *in vivo* was developed (GEHRING, 1970; CHAN and GEHRING, 1971). In order to test blastoderm cells for their developmental capacities, cells from anterior and posterior half embryos were dissociated, intermixed separately with genetically marked cells from whole blastoderms (Fig. 4), reaggregated and cultured *in vivo*. The results are summarized in Table 1, which lists the number of test implants in which a particular structure was found. The structures are divided into groups according to their disk of origin. The cells from whole blastoderms represent an internal control and give rise to epidermal structures from all regions of the fly. The cells from anterior half embryos form head and thoracic structures only,

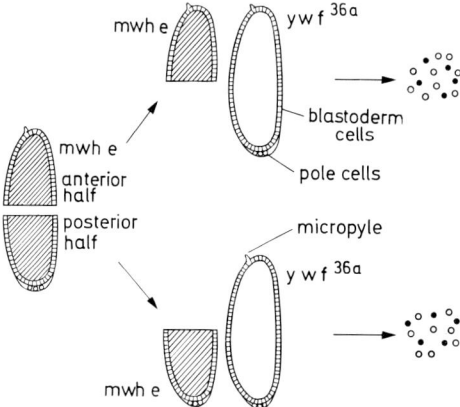

Fig. 4. Experimental procedure for analyzing the state of determination of blastoderm cells. Genetically marked blastoderm embryos are cut into half and separately intermixed with whole embryos carrying different genetic markers. Subsequently the embryos are dissociated into single cells, reaggregated, and cultured *in vivo*. Marker genes: *mwh multiple wing hairs, e ebony, y yellow, w white, f^{36a} forked* (After CHAN and GEHRING, 1971)

whereas the cells from the posterior half are restricted for forming thoracic and abdominal structures. This result clearly indicates that at least a general determination for anterior and posterior epidermal structures occurs at the time of blastoderm formation. Observations on the recessive mutant syndrome "bicaudal" also suggest an initial determination for head, thorax and abdomen (BULL, 1966). Homozygous "bicaudal" females lay defective eggs which give rise to embryos consisting of two abdomina arranged in mirror-image symmetry, irrespective of the genotype of the sperm. This "maternal effect" indicates the presence of determinative factors in the egg cytoplasm. Whether cell determination is already more specific at the blastoderm stage is an open question. Phenocopy experiments suggest an early determination of the thoracic segments. Treatment of wildtype embryos with ether or a heat shock produces flies in which the metathorax with its halteres is transformed into a second mesothorax with wings (HENKE and MAAS, 1946; GLOOR, 1947). These treated flies show the same phenotype as *bithorax* (BRIDGES and MORGAN, 1923) and *tetraptera* (ASTAUROFF, 1929) mutants. Since the highest frequency of transformed flies was obtained at the time of blastoderm formation, this might suggest a segment specific determination at the blastoderm stage, but there is no direct evidence for this interpretation. From the gynandromorph studies mentioned above, an estimate of the number of progenitor cells giving rise to a disk, can be made (STERN, 1940; GARCIA-BELLIDO and MERRIAM, 1969; BRYANT and SCHNEIDERMAN, 1969; BRYANT, 1970).

Table 1. Developmental Capacities of Genetically Marked Blastoderm Cells Isolated from Anterior and Posterior Half Embryos and Intermixed with Cells from Whole Embryos (After CHAN and GEHRING, 1971)

	Number of embryos	Total number of test-implants	Head structures[a]				Thorax structures				Abdominal structures	
			Labial disk	Clypeo-labral "disk"	Antennal disk	Eye disk	Leg disk	Wing disk	Haltere disk		Abdominal "histoblasts"	Genital disk
Anterior halves	992	801	2	3	10	4	7	16	2		1?	0
Whole embryos	470		5	0	13	5	23	35	4		21	4
Posterior halves	1471	1272	0	0	0	0	1	7	1		16	1
Whole embryos	890		6	1	5	3	15	15	8		34	2

[a] The adult structures are arranged according to their disk of origin, and the number of test-implants containing respective structures are given.

The estimates range from about 2—30 cells for the various disks. Clones of genetically marked cells induced by X-irradiation of blastoderm embryos are confined to an area of epidermis which is derived from one disk only, for example the leg disk (BRYANT and SCHNEIDERMANN, 1969; BRYANT, 1970). This finding suggests that the prospective leg disk cells are segregated into a group of their own. However, we cannot draw any safe conclusions about their determination, since we do not know how the progenitor cells for the various disks are spaced. In the case of the female genital disk, separate progenitor cells for anal plates and vaginal plates seem to be present as early as the blastoderm stage (ULRICH, 1971). By means of UV irradiation of later embryonic stages GEIGY (1931) was able to induce specific defects in certain structures of the adult fly, whereas the larval structures remained unaffected. Flies were obtained, for example, with one leg or wing missing, or with duplications of leg segments. Since in these cases the larval organs apparently were normal, we can assume that the primordia for the adult structures are already separated from the larval tissues during embryogenesis, even before the imaginal disks can be morphologically detected.

During larval development the number of cell types to which a disk cell can give rise is further restricted. For example, certain cells of the eye disk of a late second instar larva still have two possible pathways of differentiation; they may either differentiate into ommatidia or into cells forming the border of the eye (BECKER, 1957). Clones induced at later stages are confined to either one of these structures, but do not contain both of them. Similarly, certain cells of the male genital disk in the second or even early third larval instar may still give rise to both clasper and lateral plate cells (ULRICH and NÖTHIGER, unpublished). Towards the end of the third larval instar the cells become further determined for one of the alternative pathways. At this stage the method of following the cell lineage becomes unapplicable, since the marked cells divide only a few times before metamorphosis and therefore produce only a tiny patch of marked tissue.

During the last (third) larval instar the disks are most accessible to embryological surgery. By cutting the disk from a mature larva into definite fragments which are then transplanted separately into host larvae of the same age, it can be shown that different regions of the disk form different organs. The anlagen for the different organs can in this way be mapped. The most detailed maps have been established for the male and female genital disk (HADORN, BERTANI, and GALLERA, 1949; URSPRUNG, 1959), the wing disk (HADORN and BUCK, 1962) and the male foreleg disk (SCHUBIGER, 1968). Another method consists in UV microbeam treatment to eliminate small limited areas of the disks for a more refined analysis (URSPRUNG, 1957 and 1959).

Even within an individual organ anlage, further details can be mapped. For example, a small sense organ in the third antennal segment, the sacculus, is always derived from the anterior half of the antennal disk which contains only a part of the anlage for the third antennal segment (GEHRING, 1966a). In the trochanter anlage of the male foreleg disk, a region determined for forming a single bristle has been mapped (NÖTHIGER and SCHUBIGER, 1966; SCHUBIGER, 1968). These results indicate that determination of the imaginal disks occurs in a stepwise manner, progressively restricting the differentiation potential of the cells.

Although the fragmentation experiments show a considerable degree of "intradisk" determination, they give no information about the *state of determination of an individual cell.* To study this problem, isolated cells obtained by trypsinization or mechanical dissociation of disks carrying different marker genes were intermixed, reaggregated and tested in a metamorphosing host for their differentiative capacities (HADORN, ANDERS, and URSPRUNG, 1959; URSPRUNG and HADORN, 1962; NÖTHIGER, 1964; TOBLER, 1966; GARCIA-BELLIDO, 1966a and 1966b). The reaggregates form organized adult structures which may be composed of cells of one genotype only, or cells of both genotypes forming a genetically mosaic structure. *Drosophila* offers a number of convenient marker genes: *ebony* and *yellow* mutants which produce dark and yellow color, respectively, of the chitinous structures, the *singed* mutant which is characterized by altered bristle shape; the *multiple wing hairs* mutant in which the single cell hairs (trichomes) are replaced by groups of trichomes. Since both bristle shafts and trichomes are each produced by single cells, the differentiation of a single cell can be traced in this way.

In dissociation-reaggregation experiments, mosaic structures composed of cells with different genotypes are formed only by *isotypic* cells (NÖTHIGER, 1964). For example, anal plate cells will frequently form mosaics with anal plate cells of a different genotype or sex, but fail to associate with clasper cells. *Heterotypic* cells tend to separate from each other. From this result we conclude that the presumptive anal plate cells differ specifically from the clasper cells. The association of isotypic cells and the separation of heterotypic cells is most likely achieved by cell migration and selective adhesion of the cells, as observed in vertebrate cells, since there is no indication of selective cell death.

If cells from different disks having no isotypic cells in common are intermixed, the cells sort out completely and do not form mosaic structures. The only exceptions to that rule are the so-called "faulty" mosaics which occur very rarely (NÖTHIGER, 1964). If, for example, a single cell of a genital disk (or a very small group of cells) is trapped in a large area of wing disk cells, it differentiates autonomously and forms a specific genital bristle. Thus, the surrounding blastema does not exert a determinative influence upon the isolated cell. This shows that the quality (specificity) of determination at this stage is carried by the individual cell rather than by the blastema as a whole.

Are the cells of a disk from a mature larva fully programmed, or does further determination occur during the pupal stage? There is some evidence for further determination at the early pupal stage during which cell division continues. In the case of the bristle which consists of two external cells — the shaft cell (trichogen) and the socket cell (tormogen) — the external cells are presumably derived from a bristle mother cell by a differentiative division at this late stage (LEES and WADDINGTON, 1942).

Therefore, we are inclined to assume a prefinal determination which leaves the cells a limited array of possible pathways. Furthermore, there is also an indication of cell interactions leading to the determination of a specific cell type after pupation (TOBLER, 1966). Certain leg and wing bristles are accompanied by a special trichome called a bract (HANNAH-ALAVA, 1958; PEYER and HADORN, 1965), occupying a definite position in relation to the bristle. The cell which forms a bract is not derived from the bristle mother cell, since genetic mosaics are observed in which the bract cell differs in its genotype from the bristle cell. Isolated bracts without a corresponding bristle are never observed after reaggregation of isolated cells from mature larvae, which may indicate that the bristle induces the formation of a bract in a neighbouring cell after pupation (TOBLER, 1966). More recently TOBLER was able to induce the formation of bristles without sockets by treating imaginal disks with mitomycin C or nitrogen mustard (TOBLER, 1969; TOBLER and PFLUGER, 1970; TOBLER and MAIER, 1970). Such bristles lack both sockets and bracts, which perhaps suggests that the presence of either the socket cell or the complete bristle organ is a necessary prerequisite for bract formation. Alternatively, the drug might affect sockets and bracts independently without inhibiting the formation of the shaft.

In summary the present evidence indicates that the individual cells of an imaginal disk of a mature larva are determined to form one or at most very few specific cell types.

IV. Mirror-Image Duplication and Regeneration

The determination of primordia within a disk was clearly demonstrated in the fragmentation experiments mentioned above. In those experiments fragments of disks from mature donors were implanted into larvae of the same age, which forces the implant to undergo metamorphosis immediately. The fragment then forms just those structures for which it is determined.

Now we would like to know if, when metamorphosis is delayed to allow further proliferation, a fragment can form a complete set of structures which are normally derived from the entire disk. Metamorphosis can be delayed by transplanting the fragment into a younger host larva or by a two-step experiment, in which the fragment is first cultured in an adult host and then transplanted back into a larva. In both cases the implant proliferates further before differentiation even though it is derived from a disk of a mature larva. The interpretation of the results of such experiments was obscured for a long time by the fact that they were performed on the genital disk which is bilaterally symmetrical; furthermore the complete separation of the different organ anlagen of the genital disk is very difficult. By using an asymmetrical paired disk, such as the wing disk, the following results were obtained which are easier to interpret and which apply to the genital disk as well. According to the anlage map of the wing disk (HADORN and BUCK, 1962) the proximal portion of the disk contains the anlagen for the dorsal mesothorax, whereas the distal fragment forms the wing. Proximal fragments which are determined for thorax structures were tested for their ability to "regenerate" wing structures after a prolonged period of proliferation (GEHRING, 1966a). All the fragments analyzed showed additional thorax structures, but in no case was regeneration of wing structures observed. This leads to the conclusion that the quality of determination is generally reproduced by the anlagen during proliferation.

Similar results were obtained with fragments of antennal disks. If an anterior fragment of a mature antennal disk, which is determined to form one palpus, is transplanted into a younger host larva it forms two symmetrical palps of normal size and does not regenerate structures derived from the posterior region of the disk (GEHRING, 1966a). These findings have been confirmed for the leg disk by fragmentation experiments and also for the genital disk by the sophisticated method of using UV microbeam treatment to eliminate certain organ anlagen of the disk (NÖTHIGER and SCHUBIGER, 1966). We therefore conclude that the fragments of disks show a mosaic development even after prolonged proliferation and give rise only to those structures for which they are determined. This interpretation obtained further support by the cell lineage studies on the female genital disk by ULRICH (1971) who showed that in the course of duplication anal plate cells only give rise to anal plate cells, and vaginal plate cells also retain their determination.

The structures which are formed during proliferation are not just oversized organs but represent bilaterally symmetrical duplications of normal sized organs (GEHRING, 1966a; WILDERMUTH, 1968). This implies a capacity of the disks for regulating size and symmetry of the organs. Since this mechanism involves proliferation, the term *proliferative regulation* (NÖTHIGER and SCHUBIGER, 1966) has been proposed. Its mechanism, however, remains to be elucidated.

Although mirror-image duplications are the general rule, regeneration of missing parts has been reported in several instances. Regeneration was found occasionally in cultures of antennal disks ("regenerative Neubildungen", GEHRING, 1966a), but SCHUBIGER (1971) was able to show that regeneration of a complete leg from fragments of leg disks occurs regularly, if a certain area of the disk is included in the fragment. Otherwise mirror-image duplications of the pre-existing anlagen are formed. Similarly, it can be shown that the eye region of the eye-antennal disk is capable of regenerating a complete antenna, whereas the antennal disk never was found to give rise to eye structures (GEHRING, unpublished). In both cases regeneration seems to occur in a proximo-distal direction, but no such proximo-distal regeneration was found for the wing disk as mentioned above.

V. Cell Heredity and Transdetermination

By serial transplantation of imaginal blastemas into adult hosts (see p. 38) the phase of cell proliferation which is characteristic of the larval stage can be prolonged indefinitely. Samples which are transplanted back into larvae after each passage in an adult host (called a transfer generation) provide information about the capacity of the culture for differentiation. Since the original blastema is determined, the question arises whether the quality of determination is replicated in the cultures. In long-term cultures in adult hosts a continuous replication of the original determination qualities is in fact observed. In certain sublines of cultures of male genital disks, cells determined for anal plate formation were continuously reproduced during more than 70 transfer generations over a period of several years (HADORN, 1967). It was found that the determination quality is continuously transmitted from one transfer generation to the next. Since the individual cell rather than the blastema as a whole carries the quality of determination (p. 43) we designate this process as *cell heredity*. In certain sublines a loss of the capacity for anal plate formation is observed. This is due to the fact that the cultured blastema is a mosaic of organ anlagen. As the cultured blastema is cut at random into a number of fragments at each transfer to a new adult host, some of the fragments may not contain any cells determined for anal plate formation and therefore give rise to sublines which have lost the capacity to form anal plates.

However, in the cultured blastemas occasional *changes in cell heredity* affecting determination occur, which were first observed by SCHLÄPFER (1963) and HADORN (1963). For example, cultured genital disks can give rise not only to genital structures (*autotypic elements*) but also to antennal and leg structures which in normal develop-

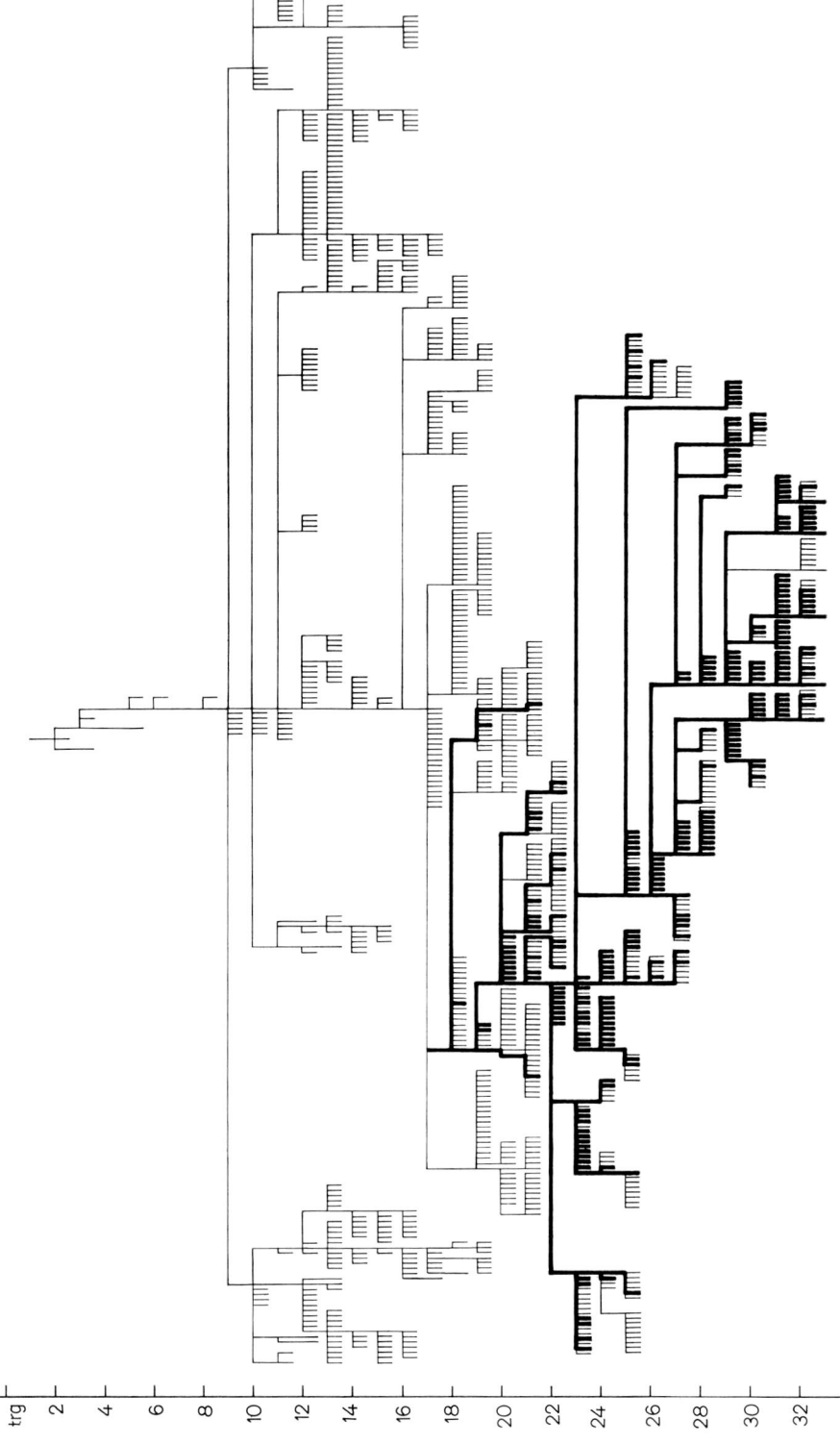

ment are formed by other disks *(allotypic elements)*. The process leading to such allotypic structures is designated as *transdetermination* (HADORN, 1965). Between the autotypic and the allotypic structures, a sharp borderline can be recognized; there is no transition zone in which the cells would show some sort of intermediate differentiation with characteristics from both autotypic and allotypic structures. Thus, the different development pathways appear to be mutually exclusive. This phenomenon has been designated as "canalization of development" (WADDINGTON, 1956).

By means of marker genes it can be shown that the allotypic cells are derived from the implant and not from invading host cells (GEHRING, 1966a). When antennal disks carrying the marker gene *yellow* are cultured in hosts homozygous for *ebony* and *multiple wing hairs*, the allotypic structures always show the donor genotype. Furthermore, in cultures of male genital disks grown in female hosts only, allotypic tarsal structures with male sexcombs were observed and all the cultured cells showed a male karyotype (HADORN, 1966; REMENSBERGER, 1968). Thus, a contribution from host cells can be ruled out. However, the possibility could not be excluded that the allotypic cells are derived from undetermined cells contained in the blastema rather than from determined autotypic cells which underwent transdetermination. This possibility will be discussed in the following section.

The information for allotypic differentiation is "inherited" in the same way as the information for autotypic differentiation. In a "pedigree" of the test implants (p. 38) the continuity of cell heredity is most obvious in those allotypic structures which are rarely formed by transdetermination. An example is illustrated in Fig. 5, which represents the pedigree of the test implants derived from a culture of haltere disks (GEHRING, MINDEK, and HADORN, 1968). The autotypic haltere elements have a slow rate of proliferation and were therefore lost after 12 transfer generations. The first transdetermination led from haltere-forming cells to presumptive wing cells, which were followed by other allotypic structures not indicated in the pedigree. In the 18th transfer generation, the wing-forming cells gave rise to presumptive eye-cells (dark lines in Fig. 5). This transdetermination step occurs very rarely. After the initial transdetermination the determination for eye-formation is continuously inherited over more than 14 transfer generations and thousands of ommatidia are formed.

VI. Clonal Analysis

The problem whether the allotypic cells are derived from undetermined cells or from determined autotypic cells can be resolved by an analysis of clones derived from single disk cells. However, at the present time we are not able to culture single cells which would form a differentiating clone. This difficulty was overcome by the application of genetic methods (GEHRING, 1967). The procedure is illustrated in Fig. 6. Larvae heterozygous for several marker genes were treated with X-rays in order to

---

Fig. 5. Cell heredity of the eye-forming cells in a long-term culture initiated by haltere blastemas. "Pedigree" of the test implants derived from the culture during 32 transfer generations (trg). Each test implant is represented by a short vertical line. In the 18th generation transdetermination led to eye-forming cells (dark lines) which reproduced their quality of determination over more than 14 transfer generations

induce genetically marked single cells in the imaginal disks by somatic crossing-over or by somatic mutation giving rise to homozygous or hemizygous cells (cf. STERN, 1936; BECKER, 1957). The antennal disks from the irradiated larvae were cultured in adult hosts, where the marked single cells give rise to a *clone*. The clones can be

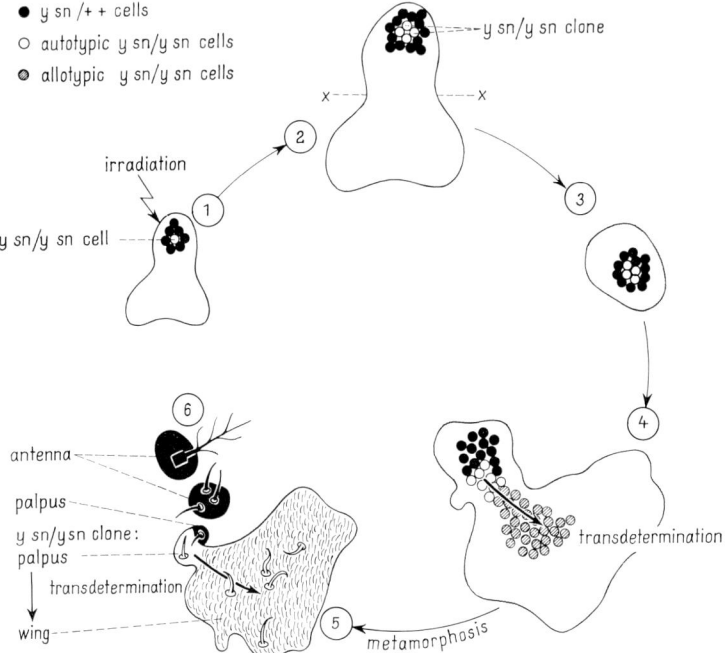

Fig. 6. Induction of clones derived from a genetically marked single cell in cultures of antennal disks (after GEHRING, 1967). 1. X-ray treatment of second instar larvae heterozygous for the recessive marker genes *yellow (y)* and *singed (sn)* and induction of a *y sn*-marked cell *(y sn/y sn)* by somatic crossing-over in the eye-antennal disk. 2. Dissection of the antennal disk from the mature larva. 3. Transplantation into an adult host. 4. Proliferation of the cultured disk and formation of a clone of *y sn/y sn* cells in which transdetermination occurs. 5. Transplantation into a larval host with which the cultured disk undergoes metamorphosis. 6. Metamorphosed structures including heterozygous autotypic cells *(y sn/+ +)* and a clone of *y sn/y sn* cells. The clone consists of autotypic palpus cells and allotypic wing cells. (This particular case is illustrated in Fig. 8 in detail)

detected after transplantation back into a larval host. The frequency of somatic crossing-over is sufficiently low that the probability of two independent crossing-overs in one organ anlage is very small. According to our calculations the observed marked areas are therefore derived from a single cell. From a total of 94 marked clones, 78 contained autotypic cells only. The remaining 16 clones listed in Fig. 7 mostly contained both autotypic and allotypic cells. A representative clone of *yellow singed (y sn)* marked cells is illustrated in Fig. 8. The autotypic antennal parts are not marked and show wild-type coloration (+). The palpus, which is also an autotypic element formed by the antennal disk, is a genetic mosaic of + and *y sn* areas

Stability of the Determined State in Cultures of Imaginal Disks in Drosophila 49

with twisted yellow bristles. The *y sn* marked base of the palpus is in direct connection with a large allotypic wing area which shows the same phenotype *(y sn)*. Thus, the

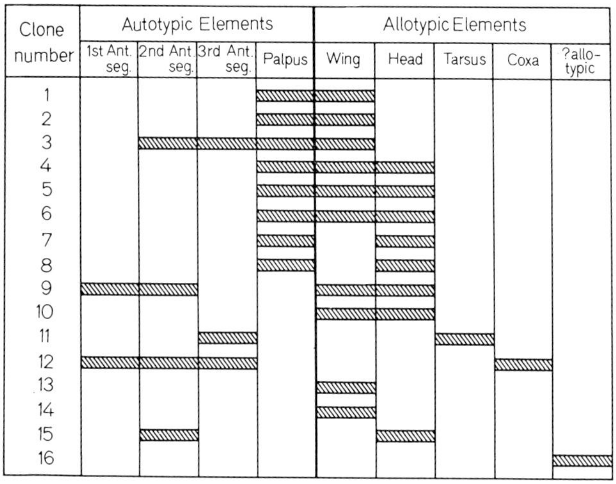

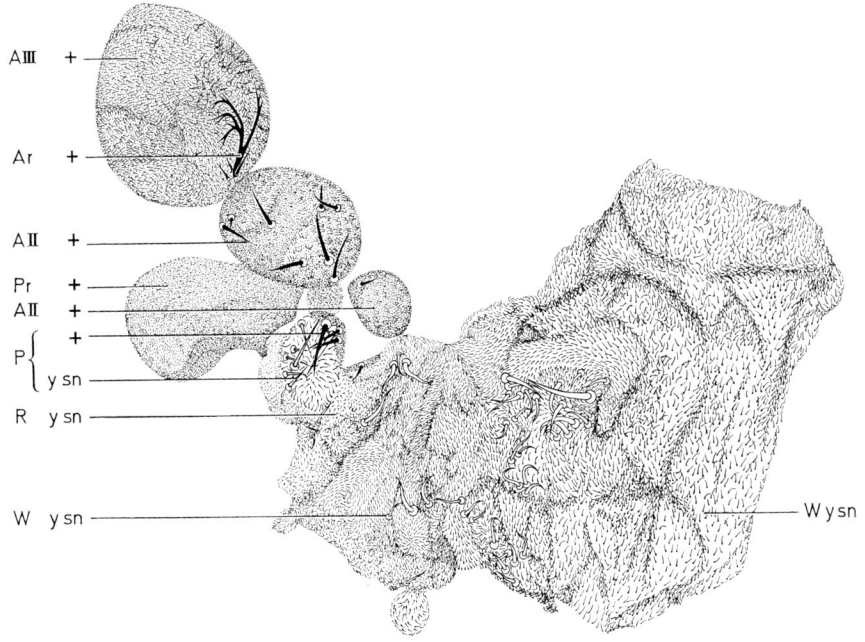

Fig. 7. Composition of 16 different clones containing allotypic cells in cultures of antennal disks (after GEHRING, 1967). cross-hatched bars: structures contained in the clone. Ant. seg. antennal segment

Fig. 8. Transdetermination in a clone of *y sn*-marked cells derived from a culture of antennal disks (after GEHRING, 1967). The clone includes part of the autotypic palpus *(P y sn)* and the rostral membrane *(R y sn)* which gave rise to the allotypic wing *(W y sn)* by transdetermination. The remaining autotypic parts designated with + show wild-type coloration. *A* II, *A* III second and third antennal segment, *Pr* prefrons (Magnification 50×)

clone includes part of the palpus region and the whole allotypic wing area. This pattern can only be explained by assuming that a somatic crossing-over or mutation occurred in an autotypic cell of the palpus anlage which subsequently gave rise to the other autotypic and allotypic cells of the clone. In five clones a similar pattern of palpus and wing cells was observed. In two clones the autotypic palpus cell had multiplied considerably and formed several palps. This indicates that the clone mother cell was indeed determined for palpus since it transmitted its quality of determination to its descendents over many cell generations. Therefore, we conclude that *transdetermination can occur in clones derived from a determined autotypic cell*. It must be emphasized that although these clones originate from a single autotypic cell, the transdetermination might not have occurred in a single cell of the clone but could have involved a simultaneous response of several cells (see p. 51).

Transdetermination is a directed process leading from one cell type to a limited array of other cell types. For example, in the clone discussed above, palpus cells gave rise to specific wing cells. The observed *sequences of transdetermination* in the clones listed in Fig. 7 are summarized in the scheme of Fig. 9a. The step from palpus to antenna involves two kinds of autotypic cells both derived from the antennal disk and is therefore designated as *region specific transdetermination* (GEHRING, 1966a). A more comprehensive scheme based upon the work of several authors on various disks (see HADORN, 1966) is given in Fig. 9b. This scheme is formulated in more general terms, since the evidence for the various steps is rather indirect. For example, haltere stands for all the different cell types constituting the haltere disk, since we do not know yet which cell type undergoes transdetermination to wing-forming cells. As the wing cells arise in cultures of haltere disks before any other allotypic cells are observed (GEHRING, MINDEK, and HADORN, 1968), we conclude that the wing cells are derived directly from haltere cells.

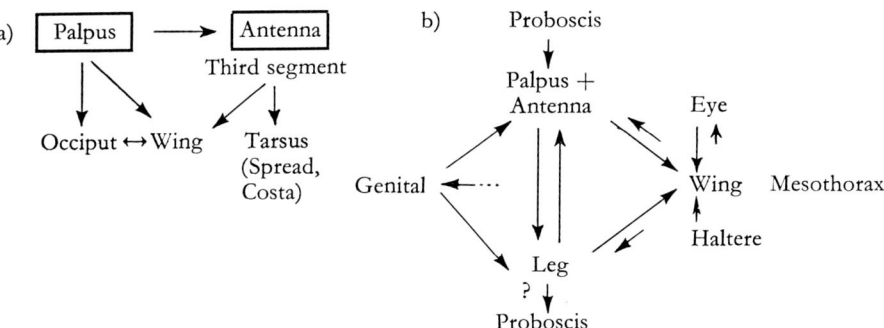

Fig. 9. Schemes of the transdetermination sequences. a Scheme based on clonal analysis. ⬜ autotypic elements. b Comprehensive scheme summarizing the observations of several authors. Short arrows indicate rare transdeterminations. The dotted arrow indicates that transdetermination to genital cells was observed but it is not known from which cell type these are derived

The scheme in Fig. 9b shows that certain transdetermination steps are *reversible* as was first shown for the change from the third antennal segment to tarsus (GEHRING, 1966a, and SCHUBIGER, 1968). Other steps seem to be irreversible. For example, no transdetermination leading to haltere structures has been observed. Thus, certain transdeterminations occur relatively frequently whereas others occur very rarely if ever. The scheme in Fig. 9 apparently does not show any correlation with the arrangement of the various organs *in situ*. The only case indicating such a correlation is the transdetermination from female genital disks to antennal segments, which seem to arise in the same proximo-distal order in which they are arranged *in situ* (MINDEK, 1968).

The *relative frequency* of transdetermination in repeated experiments remains constant (see Table 2). The only significant difference between the two experiments listed, is due to X-irradiation in the second experiment which induced a higher frequency of tarsal structures also in the controls (see p. 53). Thus, each transdetermination from one cell type to another occurs with a definite probability.

Table 2. Relative frequencies of the different allotypic elements in short-term cultures of antennal disks in independent experiments. The relative frequencies are expressed with respect to the total number (N) of cultures containing allotypic elements

Allotypic element	Relative frequency	
	1st experiment	2nd experiment
	(N = 54)	(N = 118)
Wing	0.87	0.81
Thorax	0.13	0.17
Tarsus	0.24	0.49[a]
Proximal leg segments	0.06	0.09
Head	0.44	0.58
Cibarium	0	0.06

[a] The increased frequency of tarsal structures is due to irradiation in the second experiment.

An important problem which can be solved using genetic mosaics is whether a transdetermination occurs independently in individual cells or simultaneously in several neighbouring cells. In our cloning experiments we found several cases in which the allotypic areas were genetic mosaics. However, this finding cannot be considered as evidence for a simultaneous transdetermination in both a marked and a wild-type cell, since isotypic cells tend so associate (p. 43) and the mosaic therefore might be formed by two independent transdeterminations followed by association of the allotypic cells. Nevertheless, if the border line between the areas of marked and wild-type cells in the contact zone between two different organs is carefully analyzed, it is obvious that the distribution of the cells is non-random. An example is illustrated in Fig. 10 which shows the contact zone between two different allotypic structures, an occiput and a wing area. In this case the cells are marked with the gene *multiple wing hairs (mwh)* which leads to the formation of a group of smaller trichomes (cell hairs) instead of a single larger one. The *mwh* cells form a continuous area across the border between the occiput and the wing area. This pattern cannot be explained by random association of cells arising from two independent transdeterminations. Therefore, the same kind of *transdetermination* must have occurred *synchronously* in at least two neighbouring cells, an *mwh* and a wild-type cell. This conclusion was confirmed by HADORN, GSELL, and SCHULTZ (1970) and GSELL (1971) who analyzed clones of cells marked by position-effect variegation: Transdetermination in autotypic mosaic blastemas can give rise to mosaic allotypic structures, which again implies that transdetermination occurs in groups of cells rather then in a single cell.

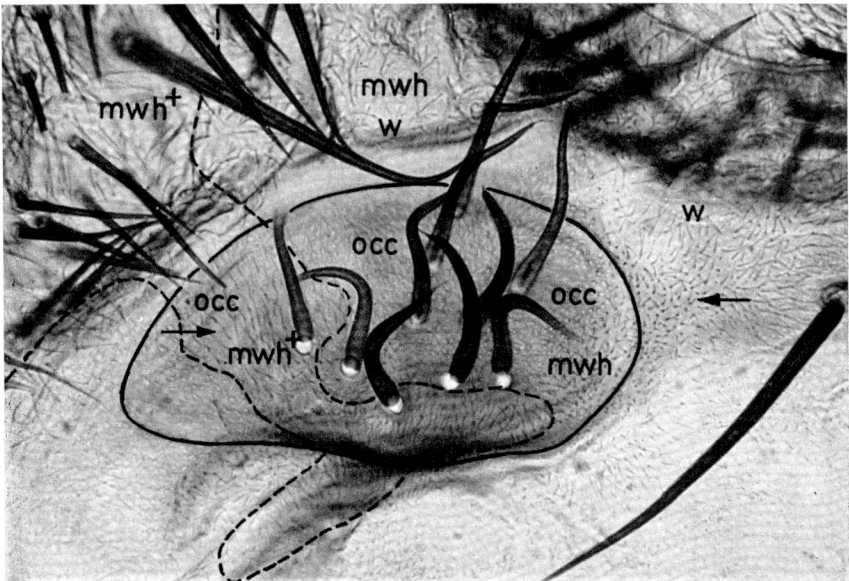

Fig. 10. Clone of genetically marked *(mwh)* cells extending over the border between head (occiput) and wing structures (after GEHRING, 1967). In the mutant *multiple wing hairs (mwh)* the cells form groups of shorter trichomes (cell hairs) in a disorderly arrangement whereas the wild-type cells *(mwh$^+$)* have larger single trichomes pointing in the same direction (arrows). The *mwh*-marked cells form a continuous area across the border between the occiput (occ, and the wing (w). This indicates that transdetermination from presumptive occiput to wing cells (or *vice versa*) takes place synchronously in at least two neighbouring cells, a *mwh*-marked and a wild-type cell. ——— border line between head (occiput) and wing structures; - - - - - border line between *mwh* and wild-type (+) cells. (Magnification 500 ×)

VII. Possible Causes of Transdetermination

At first thought the *culture medium*, i.e. the adult hemolymph, might be a possible cause of transdetermination. Yet, this possibility is very unlikely since transdetermination occurs both in larval and adult hosts (HADORN, 1963; GEHRING, 1966a). It was also shown that the frequency of transdetermination does not depend directly on the period of time the blastema is cultured (TOBLER, 1966). The only factor so far detected which has an influence on the frequency of transdetermination is proliferation. First, a highly positive correlation between the frequency of transdetermination and the extent of proliferation was observed (TOBLER, 1966). However, a correlation does not necessarily imply a causal relationship. There is another observation of importance in this connection. If a labial disk is cultured *in vivo* it usually duplicates and forms a complete symmetrical proboscis (WILDERMUTH and HADORN, 1965; WILDERMUTH, 1968a). By labelling with thymidine and autoradiography it was shown that replication takes place all over the disk, but the rate of cell division seems to be much lower on the original side of the disk than on the opposite side formed by duplication (WILDERMUTH, 1968b)[2]. If transdetermination occurs, the allotypic cells are always found on the "newly formed" side which underwent more replication cycles but is otherwise identical. This observation strongly suggests that *cell divisions are a necessary prerequisite* for transdetermination. The actual cause of transdetermination is still unknown. It seems to reside in the mechanism of cell heredity itself, which sometimes fails to replicate the quality of determination and thereby causes a transdetermination. A hypothetical scheme for intrinsic cellular mechanisms which might lead to transdetermination was proposed by HADORN (1967). It is mainly based upon the observation that cell division is a necessary prerequisite for transdetermination. A higher rate of cell division presumably causes a dilution of the carriers of determination which might lead to the activation of other sets of genes by a feedback system. However, there is no direct evidence for a higher rate of proliferation of those cells undergoing transdetermination as compared to other cells in the culture. Furthermore, dilution of the carriers alone would not explain the fact that certain transdeterminations are reversible. The alternative differentiation into mutually exclusive pathways might be explained by assuming one or a very small number of controlling genes regulating integrated sets of genes responsible for the differentiation of an organ (cf. GEHRING, 1966a).

Developmental changes leading to the same effect as transdetermination can be produced by mutations, the so-called *homeotic mutations* which have been known for a long time (for discussion of the genetic analysis see LEWIS, 1964; GEHRING and NÖTHIGER, in press). These mutations may interfere drastically with determination. Thus, in the mutant *Nasobemia* which appears to be a single gene mutant, the antenna and the adjacent part of the head capsule are transformed into a complete middle leg and the corresponding part of the ventral thorax, which normally are derived both from the same leg disk (GEHRING, 1966b and 1970). This shows that a single gene mutation can bring into action all the genes necessary for the differentiation of a leg disk in a blastema which would otherwise form head structures. The mutant *aristapedia* (BALKASCHINA, 1929) has a more limited effect in that it transforms only the arista and the adjacent part of the third antennal segment into a tarsus. When *aristapedia* antennal disks are cultured and tested in wild-type hosts, tarsal structures are found (GEHRING, 1966a). However, after a few transfer generations aristae are also found although this particular mutant (ss^a) never forms an arista *in situ*. This indicates that the mutant has not lost the genetic information to produce an arista. Probably the mutation affects a mechanism controlling which of the two alternate determination programs will be expressed.

From several of these mutants *phenocopies* may be obtained by various treatments. For example, phenocopies of the mutant *aristapedia* are produced by treating the larvae with nitrogen mustard (BODENSTEIN and ABDEL-MALEK, 1949), sodium metaborate (SANG and MCDONALD, 1954), 5-fluoro-uracil or X-rays (GEHRING, 1964 and 1967). All these agents seem to have one feature in common: they cause cell damage and thereby might induce additional cell divisions as a response to wounding; but their precise mode of action on the antennal disk is not known. The frequency of

---

[2] The clonal analysis of duplication of the female genital disk by ULRICH (1971) also provides evidence for a more extensive proliferation on the "newly formed" side.

phenocopies obtained varies greatly among different strains, and therefore depends to a great extent on the genetic consitution and the possible presence of subthreshold mutants. Certain strains do not respond to the treatment at all, which indicates that these phenocopies do not correspond exactly to transdeterminations, since these have been observed in all the strains investigated so far.

As changes of determination are produced by single gene mutations one might assume that transdetermination is due to *somatic mutation*. However, several arguments point against this hypothesis. Although exact data on the absolute frequency of transdetermination for the individual cell are not available, it is evident that this frequency is significantly higher than the frequency of spontaneous somatic mutations of marker genes which is very low in the cultures. Furthermore, transdetermination is a directed process leading from one cell type to another with a definite probability, and certain cell types revert with a high frequency. Finally, the strongest argument comes from the observation that neighbouring cells can undergo the same kind of transdetermination simultaneously. Thus, it seems unlikely that transdetermination is caused by somatic mutations. However, for the mutant *Antennapedia*, which also transforms part of the antenna into leg structures, it was shown that the homeotic transformation occurs in a group of cells rather than in a single cell, in much the same way as transdetermination (POSTLETHWAIT and SCHNEIDERMAN, 1969). Therefore, we are inclined to assume that homeotic mutations and transdeterminations act by a similar mechanism, except that homeotic mutations alter the gene structure, whereas transdeterminations seem to change gene activity.

VIII. Conclusions

In vivo cultures of imaginal disks provide an excellent system to maintain proliferating cells in a determined (programmed) state so that they do not undergo adult differentiation. Thus, determination can be clearly separated from the later steps of differentiation.

From observations on long-term cultures it is evident that *a specific state of determination can be propagated* over numerous cell generations and is therefore to a certain extent *stable*. There is considerable evidence that determination is carried by the individual cell rather than by the organ anlagen as a whole. For that reason the mechanism for propagation of the determined state is designated as *cell heredity*, but the carrier of determination is still unknown.

The stability of the determined state is limited. It has been shown that *changes of determination (transdeterminations)* occur in clones derived from determined cells. They lead to a new quality of determination which again can be propagated or undergo further changes. Each cell type gives rise to a limited array of other cell types with a definite probability. The majority of the observed transdeterminations are *reversible*, and completely stable blastemas which do not revert are rarely obtained. The factors causing transdetermination presumably reside in the mechanism of cell heredity. There is accumulating evidence that *cell divisions* are a *necessary prerequisite for transdetermination*. Since transdetermination is a directed and reversible process which presumably occurs in groups of contiguous cells simultaneously, we assume that it involves changes of gene activity rather than changes in the structure of the genetic material.

The main problem for future research seems to be the identification of the carrier of determination which might give the key to the whole problem of cell differentiation.

References

ASTAUROFF, B.: Studien über die erbliche Veränderung der Halteren bei *Drosophila melanogaster*. Wilhelm Roux' Arch. Entwickl.-Mech. Org. **115**, 424—447 (1929).

BALKASCHINA, E.: Ein Fall der Erbhomöosis (Die Genovariation "aristopedia") bei *Drosophila melanogaster*. Wilhelm Roux' Arch. Entwickl.-Mech. Org. **115**, 448—463 (1929).

BECKER, H.: Über Röntgenmosaikflecken und Defektmutationen am Auge von *Drosophila* und die Entwicklungsphysiologie des Auges. Z. indukt. Abstamm.- u. Vererb.-L. **88**, 333—373 (1957).

BODENSTEIN, D.: The postembryonic development of *Drosophila*. In: DEMEREC, M. (Ed.): Biology of *Drosophila*, pp. 275—367. New York: John Wiley and Sons 1950.

BODENSTEIN, D., ABDEL-MALEK, A.: The induction of *aristapedia* by nitrogen mustard in *Drosophila virilis*. J. exp. Zool. **111**, 95—115 (1949).

BRIDGES, C., MORGAN, T.: The third chromosome group of mutant characters of *Drosophila melanogaster*. Carnegie Inst. of Washington Publ. No. 327, p. 225 (1923).

BRYANT, P.: Cell lineage relationships in the imaginal wing disc of *Drosophila melanogaster*. Develop. Biol. **22**, 389—411 (1970).

BRYANT, P., SCHNEIDERMAN, H.: Cell lineage, growth and determination in the imaginal leg disc of *Drosophila melanogaster*. Develop. Biol. **20**, 263—290 (1969).

BULL, A.: *Bicaudal*, a genetic factor which affects the polarity of the embryo in *Drosophila melanogaster*. J. exp. Zool. **161**, 221—242 (1966).

CHAN, L.-N., GEHRING, W.: Determination of blastoderm cells in *Drosophila melanogaster*. Proc. nat. Acad. Sci. (Wash.) **68**, 2217—2221 (1971).

EPHRUSSI, B., BEADLE, G.: A technique of transplantation for *Drosophila*. Amer. Naturalist **70**, 218—225 (1936).

GARCIA-BELLIDO, A.: Pattern reconstruction by dissociated imaginal disk cells of *Drosophila melanogaster*. Develop. Biol. **14**, 278—306 (1966a).

GARCIA-BELLIDO, A.: Changes in selective affinity following transdetermination in imaginal disc cells of *Drosophila melanogaster*. Exp. Cell Res. **44**, 382—392 (1966b).

GARCIA-BELLIDO, A., MERRIAM, J.: Cell lineage of the imaginal discs in *Drosophila* gynandromorphs. J. exp. Zool. **170**, 61—76 (1969).

GEHRING, W.: Phenocopies produced by 5-fluoro-uracil. Drosoph. Inf. Serv. **39**, 102 (1964).

GEHRING, W.: Übertragung und Änderung der Determinationsqualitäten in Antennenscheiben-Kulturen von *Drosophila melanogaster*. J. Embryol. exp. Morph. **15**, 77—111 (1966a).

GEHRING, W.: Bildung eines vollständigen Mittelbeines mit Sternopleura in der Antennenregion bei der Mutante *Nasobemia* (*Ns*) von *Drosophila melanogaster*. Arch. Klaus-Stift. Vererb.-Forsch. **41**, 44—54 (1966b).

GEHRING, W.: Clonal analysis of determination dynamics in cultures of imaginal disks in *Drosophila melanogaster*. Develop. Biol. **16**, 438—456 (1967).

GEHRING, W.: Problems of cell determination and differentiation in *Drosophila*. In: HANLY, E. W. (Ed.): Problems in Biology: RNA in Development, pp. 231—244, 1970.

GEHRING, W., MINDEK, G., HADORN, E.: Auto- und allotypische Differenzierungen aus Blastemen der Halterenscheibe von *Drosophila melanogaster* nach Kultur *in vivo*. J. Embryol. exp. Morph. **20**, 307—318 (1968).

GEHRING, W., NÖTHIGER, R.: The imaginal discs of *Drosophila*. In: WADDINGTON, C. H., COUNCE-NICKLAS, S. (Eds.): Developmental Systems: Insects. London-New York: Academic Press (in press).

GEHRING, W., SEIPPEL, S.: Die Imaginalzellen des Clypeo-Labrums und die Bildung des Rüssels von *Drosophila melanogaster*. Rev. Suisse Zool. **74**, 589—596 (1967).

GEIGY, R.: Erzeugung rein imaginaler Defekte durch ultraviolette Eibestrahlung bei *Drosophila melanogaster*. Wilhelm Roux' Arch. Entwickl.-Mech. Org. **125**, 406—447 (1931).

GEYER-DUSZYNSKA, I.: Experiments on nuclear transplantation in *Drosophila*. Rev. Suisse Zool. **74**, 614—615 (1967).

GLOOR, H.: Phänokopie-Versuche mit Äther an *Drosophila*. Rev. Suisse Zool. **54**, 637—712 (1947).

GSELL, R.: Untersuchungen zur Stabilität einer *yellow* Positionseffekt-Variegation in Imaginalscheiben-Kulturen von *Drosophila melanogaster*. Molec. Gen. Genetics **110**, 218—237 (1971).

HADORN, E.: Differenzierungsleistungen wiederholt fragmentierter Teilstücke männlicher Genitalscheiben von *Drosophila melanogaster* nach Kultur *in vivo*. Develop. Biol. **7**, 617—629 (1963).

HADORN, E.: Problems of determination and transdetermination. Brookhaven Symp. Biol. **18**, 148—161 (1965).

HADORN, E.: Konstanz, Wechsel und Typus der Determination und Differenzierung in Zellen aus männlichen Genitalanlagen von *Drosophila melanogaster* nach Dauerkultur *in vivo*. Develop. Biol. **13**, 424—509 (1966).

HADORN, E.: Dynamics of determination. In: LOCKE, M. (Ed.): Major problems in developmental biology, pp. 85—104. New York-London: Academic Press 1967.

HADORN, E., ANDERS, G., URSPRUNG, H.: Kombinate aus teilweise dissoziierten Imaginalscheiben verschiedener Mutanten und Arten von *Drosophila*. J. exp. Zool. **142**, 159—175 (1959).

HADORN, E., BERTANI, G., GALLERA, J.: Regulationsfähigkeit und Feldorganisation der männlichen Genital-Imaginalscheibe von *Drosophila melanogaster*. Wilhelm Roux' Arch. Entwickl.-Mech. Org. **144**, 31—70 (1949).

HADORN, E., BUCK, D.: Über Entwicklungsleistungen transplantierter Teilstücke von Flügel-Imaginalscheiben von *Drosophila melanogaster*. Rev. Suisse Zool. **69**, 302—310 (1962).

HADORN, E., GSELL, R., SCHULTZ, J.: Stability of a position-effect variegation in normal and transdetermined larval blastemas from *Drosophila melanogaster*. Proc. nat. Acad. Sci. (Wash.) **65**, 633—637 (1970).

HADORN, E., HÜRLIMANN, R., MINDEK, G., SCHUBIGER, G., STAUB, M.: Entwicklungsleistungen embryonaler Blasteme von *Drosophila* nach Kultur im Adultwirt. Rev. Suisse Zool. **75**, 557—569 (1968).

HANNAH-ALAVA, A.: Morphology and chaetotaxy of the legs of *Drosophila melanogaster*. J. Morph. **103**, 281—310 (1958).

HENKE, K., MAAS, H.: Über sensible Perioden der allgemeinen Körpergliederung von *Drosophila*. Nachr. Akad. Wiss. Göttingen. Math-phys. Kl. **1**, 3—4 (1946).

ILLMENSEE, K.: Transplantation of embryonic nuclei into unfertilized eggs of *Drosophila melanogaster*. Nature (Lond.) **219**, 1268—1269 (1968).

ILLMENSEE, K.: Imaginal structures after nuclear transplantation in *Drosophila melanogaster*. Naturwissenschaften **57**, 550—551 (1970).

LEES, A., WADDINGTON, C.: The development of bristles in normal and some mutant types of *Drosophila melanogaster*. Proc. roy. Soc. B, **131**, 87—110 (1942).

LEWIS, E.: Genetic control and regulation of developmental pathways. In: LOCKE, M. (Ed.): The role of chromosomes in development, pp. 231—252 New York-London: Academic Press 1964.

MINDEK, G.: Proliferations- und Transdeterminationsleistungen der weiblichen Genital-Imaginalscheiben von *Drosophila melanogaster* nach Kultur *in vivo*. Wilhelm Roux' Arch. Entwickl.-Mech. Org. **161**, 249—280 (1968).

NÖTHIGER, R.: Differenzierungsleistungen in Kombinaten, hergestellt aus Imaginalscheiben verschiedener Arten, Geschlechter und Körpersegmente von *Drosophila*. Wilhelm Roux' Arch. Entwickl.-Mech. Org. **155**, 269—301 (1964).

NÖTHIGER, R., SCHUBIGER, G.: Developmental behaviour of fragments of symmetrical and asymmetrical imaginal discs of *Drosophila melanogaster* (*Diptera*). J. Embryol. exp. Morph. **16**, 355—368 (1966).

PEYER, B., HADORN, E.: Zum Manifestationsmuster der Mutante "*multiple wing hairs*" (*mwh*) von *Drosophila melanogaster*. Arch. Klaus-Stift. Vererb.-Forsch. **40**, 19—26 (1965).

POSTLETHWAIT, J., SCHNEIDERMAN, H.: A clonal analysis of determination in *Antennapedia* a homoeotic mutant of *Drosophila melanogaster*. Proc. nat. Acad. Sci. (Wash.) **64**, 176—183 (1969).

REMENSBERGER, P.: Cytologische und histologische Untersuchungen an Zellstämmen von *Drosophila melanogaster* nach Dauerkultur *in vivo*. Chromosoma (Berl.) **23**, 386—417 (1968).

SANG, J., MCDONALD, J.: Production of phenocopies in *Drosophila* using salts, particularly sodium metaborate. J. Genet. **52**, 392—412 (1954).

SCHLÄPFER, T.: Der Einfluß des adulten Wirtsmilieus auf die Entwicklung von larvalen Augenantennen-Imaginalscheiben von *Drosophila melanogaster*. Wilhelm Roux' Arch. Entwickl.-Mech. Org. **154**, 378—404 (1963).

SCHUBIGER, G.: Anlageplan, Determinationszustand und Transdeterminationsleistungen der männlichen Vorderbeinscheibe von *Drosophila melanogaster*. Wilhelm Roux' Arch. Entwickl.-Mech. Org. **160**, 9—40 (1968).

SCHUBIGER, G.: Regeneration, pattern duplication and transdetermination in fragments of the leg disc of *Drosophila melanogaster*. Develop. Biol. **26**, 277—295 (1971).

SCHUBIGER, G., SCHUBIGER, M., HADORN, E.: Mischungsversuche mit Keimteilen von *Drosophila melanogaster* zur Ermittlung des Determinationszustandes imaginaler Blasteme im Embryo. Wilhelm Roux' Arch. Entwickl.-Mech. Org. **163**, 33—39 (1969).

SCHUBIGER, M., SCHNEIDERMAN, H.: Nuclear transplantation in *Drosophila melanogaster*. Nature (Lond.) **230**, 185—186 (1971).

STERN, C.: Somatic crossing over and segregation in *Drosophila melanogaster*. Genetics **21**, 625—730 (1936).

STERN, C.: The prospective significance of imaginal discs in *Drosophila*. J. Morph. **67**, 107—122 (1940).

STURTEVANT, A.: The *claret* mutant type of *Drosophila simulans*: A study of chromosome elimination and of cell-lineage. Z. wiss. Zool. **135**, 323—356 (1929).

TOBLER, H.: Zellspezifische Determination und Beziehung zwischen Proliferation und Transdetermination in Bein- und Flügelprimordien von *Drosophila melanogaster*. J. Embryol. exp. Morph. **16**, 609—633 (1966).

TOBLER, H.: Beeinflussung der Borstendifferenzierung und Musterbildung durch Mitomycin bei *Drosophila melanogaster*. Experientia (Basel) **25**, 213—214 (1969).

TOBLER, H., MAIER, V.: Zur Wirkung von Senfgaslösungen auf die Differenzierung des Borstenorganes und auf die Transdeterminationsfrequenz bei *Drosophila melanogaster*. Wilhelm Roux' Arch. Entwickl.-Mech. Org. **164**, 303—312 (1970).

TOBLER, H., PFLUGER, M.: Untersuchungen zur Wirkung von Mitomycin C auf die Entwicklung der männlichen Vorderbeinscheibe und die Differenzierung des Borstenorgans von *Drosophila melanogaster* nach Transplantation in larvale Wirte. Wilhelm Roux' Arch. Entwickl.-Mech. Org. **164**, 293—302 (1970).

ULRICH, E.: Cell lineage, Determination und Regulation in der weiblichen Genitalmaginalscheibe von *Drosophila melanogaster*. Wilhelm Roux Arch. Entwickl.-Mech. Org. **167**, 64—82 (1971).

URSPRUNG, H.: Untersuchungen zum Anlagemuster der weiblichen Genitalscheibe von *Drosophila melanogaster* durch UV-Strahlenstich. Rev. Suisse Zool. **64**, 303—311 (1957).

URSPRUNG, H.: Fragmentierungs- und Bestrahlungsversuche zur Bestimmung von Determinationszustand und Anlageplan der Genitalscheiben von *Drosophila melanogaster*. Wilhelm Roux' Arch. Entwickl.-Mech. Org. **151**, 504—558 (1959).

URSPRUNG, H.: *In vivo* culture of *Drosophila* imaginal discs. In: WILT, F., WESSELLS, N., (Eds.): Methods in Developmental Biology, pp. 485—492. New York: Thomas Crowell Co. 1967.

URSPRUNG, H., HADORN, E.: Weitere Untersuchungen über Musterbildung in Kombinaten aus teilweise dissoziierten Flügel-Imaginalscheiben von *Drosophila melanogaster*. Develop. Biol. **4**, 40—66 (1962).

WADDINGTON, C.: Principles of embryology, p. 349. London: Allen & Unwin 1956.

WADDINGTON, C., PERRY, M.: The ultra-structure of the developing eye of *Drosophila*. Proc. roy. Soc. B **153**, 155—178 (1960).

WILDERMUTH, H.: Differenzierungsleistungen, Mustergliederung und Transdeterminationsmechanismen in hetero- und homoplastischen Transplantaten der Rüsselprimordien von *Drosophila*. Wilhelm Roux' Arch. Entwickl.-Mech. Org. **160**, 41—75 (1968a).

WILDERMUTH, H.: Autoradiographische Untersuchungen zum Vermehrungsmuster der Zellen in proliferierenden Rüsselprimordien von *Drosophila melanogaster*. Develop. Biol. **18**, 1—13 (1968b).

WILDERMUTH, H., HADORN, E.: Differenzierungsleistungen der Labial-Imaginalscheibe von *Drosophila melanogaster*. Rev. Suisse Zool. **72**, 686—694 (1965).

ZALOKAR, M.: Transplantation of nuclei in *Drosophila melanogaster*. Proc. nat. Acad. Sci. (Wash.) **68**, 1539—1541 (1971).

Pattern Formation in Imaginal Disks

A. García-Bellido

Instituto Genética y Antropología, Centro Investigaciones Biológicas, C.S I.C., Madrid

I. Introduction

The shape of a crystal results directly from the physical properties of its atomic elements. Probably, the shape of a virus merely depends on the physico-chemical properties of the different coat proteins, coded by the corresponding genes (see Levine, 1969). By contrast, in multicellular organisms several steps are interposed between the genetic information of the zygote and the shape of the adult organism. In this case, shape appears after cell multiplication, cell grouping and cell differentiation. Thus, any causal explanation of organic shape in multicellular organisms would need to know how the genetic information is processed to build different proteins, different cells and different cell arrangements. When looking for experimental systems to study morphogenesis, attention was soon directed to morphogenetic patterns. The term morphogenetic pattern is currently applied to shapes in which elements of different types appear located in fixed positions with respect to each other.

The integumental pattern of insects is a much favored experimental system. The elements constituting the pattern are single cell derivatives identifiable as chaetes (bristles), trichomes (hairs), and sensilla. In the adult *Drosophila*, e.g., these structures or elements vary in shape, size and color, representing several types of characteristic cell differentiations. The number of different cuticular elements identifiable in the adult integument of *Drosophila* is probably less than one hundred. The spatial arrangement of these elements is characteristic of each particular region of the body surface. However, a subdivision of the body surface in terms of separate patterns would be artificial. Some elements are unique to a given pattern, others appear in different body regions. The elements may appear singly or repetitively arranged at regular intervals, but generally different elements intermingle and arrange themselves in highly intricate patterns. However, the types of elements present in a given arrangement, and their topological distribution are constant features of patterns.

The aim of the present study is to discuss recent work, especially that in *Drosophila*[1], on the mechanics of pattern formation. We shall focus our attention to the last stages of development of the imaginal disks, when the decisions of the epidermal cells to differentiate as specific cuticular elements in given places are apparently taking place.

---

1 Pertinent reviews of the problems of pattern formation in insects are marked with an asterisk in the list of references.

II. Pattern Variations

A. Genetic Control of Patterns

The relative location as well as the number of different elements of a given pattern is highly constant in genetically related individuals. This constancy in topological arrangement and in size indicates a strong developmental control of pattern formation.

The constancy in topological arrangement permits the construction of detailed chaetotaxal maps of some body region patterns (Fig. 1 b—e). In these maps the location of the different elements can be defined in numerical terms relative to the position of neighbouring elements. Variance analysis of the number of elements of a given pattern between two wild type strains and their hybrids has shown that the genetic component of variation is higher than the environmental one (see HANNAH-ALAVA, 1958). Table 1 presents the variations in the number of elements in different patterns of both males and females of the same sibship and age. In the different patterns of both sexes the standard deviation does not exceed 10% of the mean values. Moreover, the sexual variations surpass the variations found within individuals of the same sex. This is interesting, since the sexual differences are due primarily to the genetic constitution. The variations between left and right patterns of the same individuals are lower than the variations among individuals.

Table 1. Mean number $(\bar{m})$ and standard deviation (σ) of elements appearing in some repetitive patterns. n, number of flies. Wildtype Sevelen flies of the same sibship and hatching day. Patterns described in text

Patterns	Females			Males		
	n	$\bar{m}$	σ	n	$\bar{m}$	σ
Eye, facets	39	772.0	54.5	31	732.2	60.4
Wing, medial triple row	63	94.6	3.8	54	88.6	3.6
Wing, dorsal triple row	62	22.1	1.4	54	21.1	1.5
Foreleg, last tibial transverse row	71	12.8	0.9	54	14.2	0.9
Foreleg, sexcomb	—	—	—	85	11.8	1.4
Tergites, Vth left	56	54.8	5.2	56	45.3	5.2
Genitalia, vaginal plate	56	14.1	1.4	—	—	—
Genitalia, claspers	—	—	—	50	30.5	1.8

Since the cuticular elements are single cell derivatives, the differences between individuals and sexes correspond in part to variations in the number of cells building the pattern. We have experimental evidence that the final number of cells reached by an intact imaginal disk is limited (GARCÍA-BELLIDO, 1965). Imaginal disks from newly hatched larvae cultured in the haemolymph of adult flies reach a maximum size, similar to that found in mature larvae *in situ*, irrespective of the growth conditions.

The arrangement of chaetes in the adult integument of Diptera is used for taxonomic purposes. This is an indication of their species stability. STURTEVANT (1970) compared the appearance of certain macrochaetes of the head and notum in several higher Diptera. Some of these chaetes were identified by their similar shape and position in far related species of the *Drosophilidae*. Surprisingly, a common pattern for some of these chaetes can be traced to flies in the Eocene, manifesting its constancy over 100 million generations. Variations in this pattern also exist. Some of the chaetes of the general pattern of head and notum are present or absent in different genera. Correlation coefficients between some groups of chaetes are characteristic, but combinations of different elements are present in one or another species. Interestingly, the correlations between macrochaetes and microchaetes of the same region are, in general, weak. This indicates that the actual pattern is constructed with elements, each having independent developmental determinants.

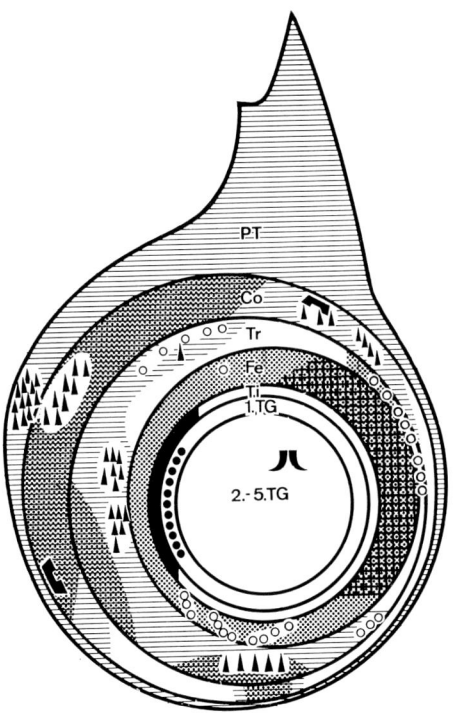

Fig. 1a

Fig. 1a—e. Anlage plan and adult structures of the leg disk. a morphogenetic map of the presumptive adult structures of the male foreleg disk of a mature larva. Symbols correspond to different kinds of single cell structures. *PT* prothorax; *Co* coxa; *Tr* trochanter; *Fe* femur; *Ti* tibia; *TG* tarsal segments (SCHUBIGER, 1968). b—e tibial and basitarsal chaetotaxy of the male b and female c foreleg, the male and female midleg d and hindleg e (HANNAH-ALAVA, 1958)

We do not know the selective mechanisms which maintained the patterns in different species and over many generations. In *Drosophila*, directed selection can increase the number of elements or suppress them, independently of other elements of the same region. Moreover, selection may uncover new patterns. Thus, SONDHI

(1962) was able to select for the appearance of a new chaete in the head, which occupied a new fixed position but left the position of the remaining neighbouring chaetes unchanged. It is interesting that the new resulting pattern was found to be typical of flies belonging to the related family of the *Aulacigasteridae*. Thus, selection in that

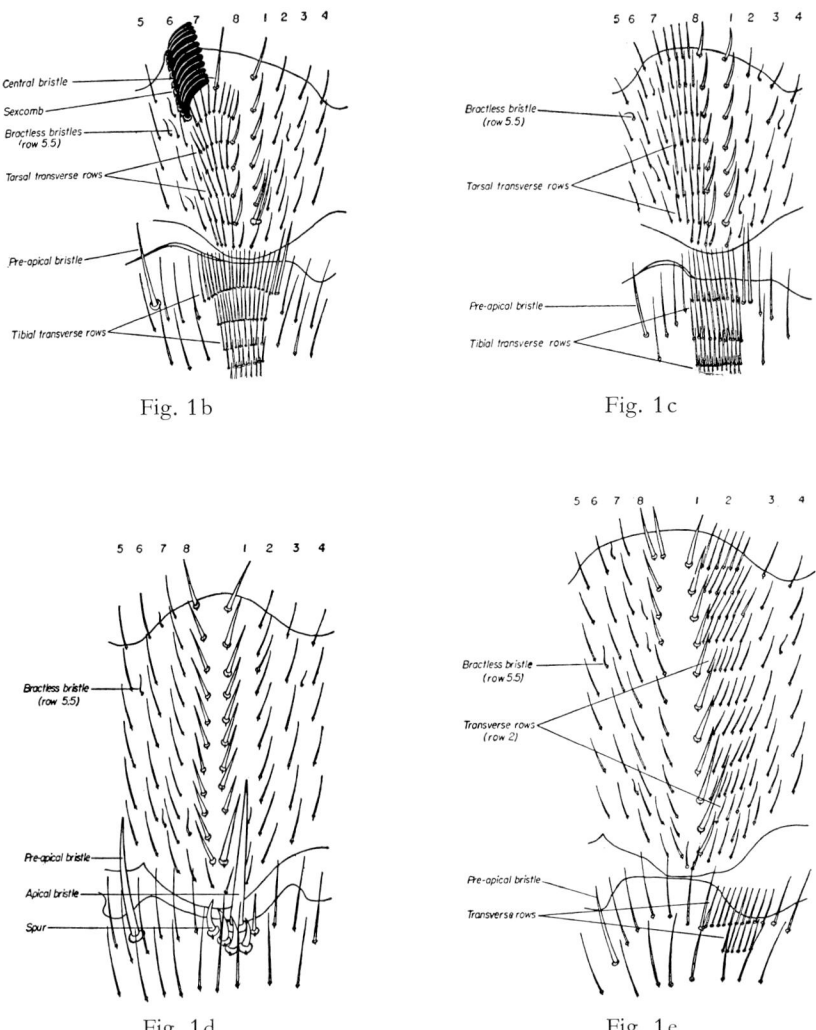

Fig. 1b

Fig. 1c

Fig. 1d

Fig. 1e

case may have unveiled a preexisting pattern. In other instances selection creates distortions of the normal chaete distribution in particular regions (MAYNARD-SMITH and SONDHI, 1960). These experiments not only emphasize the role of the genome in the control of pattern formation, but furthermore they indicate that *the location of the elements and the differentiation of the elements themselves can be independently controlled by genetic factors.*

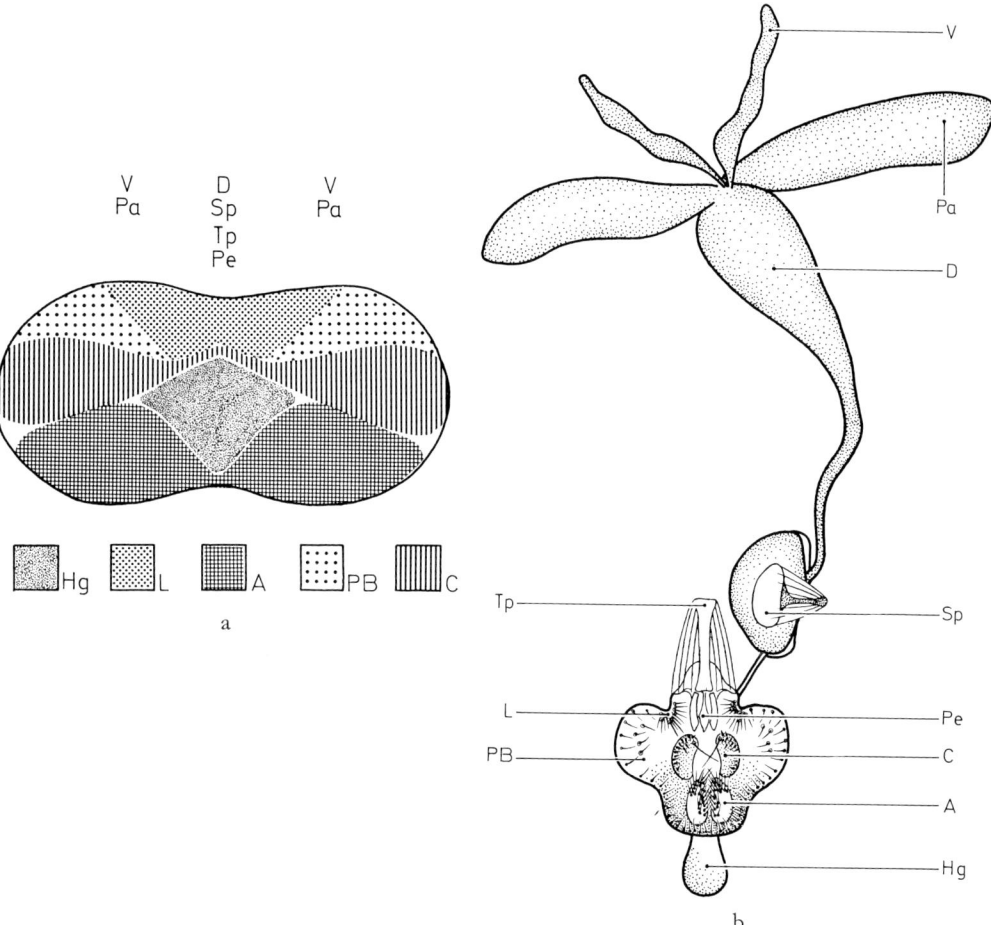

Fig. 2a, b. Anlage plan and adult structures of the male genital disk. a morphogenetic map of the presumptive adult regions (After URSPRUNG, 1959). b the corresponding adult structures (After HADORN, 1966a). *A* anal plates; *C* Claspers; *D* Ductus eiaculatorius; *Hg* hindgut; *L* Lateral plates; *Pa* paragonia; *PB* peripheral bristles; *Pe* penis apparatus; *Sp* sperm pump; *Tp* apodeme; *V* vas deferens

B. Pattern Mutants

Particular genes of *Drosophila* are known to control very specific characteristics of the adult patterns. An analysis of the different types of mutations interfering with the normal pattern can give us clues as to how the genome controls pattern formation. No comprehensive classification of known morphogenetic mutants of *Drosophila* is yet available. Here we will tentatively group some of them on the basis of the assumed nature of their pattern modifications (for detailed description of these mutants see LINDSLEY and GRELL, 1968).

Among the pattern mutants we first include those which lead to the *suppression* of entire cuticular regions more or less indiscriminately. Some dominant and recessive mutants are characterized by the deletion of eye facets, leg segments, wing regions or abnormalities in the abdomen. Some of them are known to produce cell lethality sometime during the development of the imaginal disks (see FRISTROM, 1969; SPREIJ, 1971). Interestingly, several mutations in different loci produce scalloping in different portions of the wing margin thus indicating that this cell lethality may also be region specific. Other mutants specifically suppress certain elements of the pattern leaving the rest unchanged. The classical example is that of the different alleles of *scute (sc)* (DUBININ, 1929). Each allele specifically removes some chaetes of the notum and head. Different alleles of *achaete (ac)*, a closely linked gene to *scute*, also characteristically remove other chaetes of the head and notum. In both cases the mutant pattern is allele specific, and no theoretical approach as yet has given a comprehensive explanation of their mode of action (see STURTEVANT, 1970).

A second group of mutants specifically leads to an *addition* of new elements to an otherwise unchanged pattern. *Hairy wing (Hw)*, closely linked to both *achaete* and *scute*, determines the appearance of extra chaetes in the head, notum and wing surface. The extrachaetes in the wing surface are specifically placed along certain wing veins. The recessive mutant *hairy (h)*, located in another chromosome, mimics the *Hairy wing* phenotype. Thus, the function of the normal allele of *hairy* could consist in suppressing an already preexisting pattern. As seen before, selection may also uncover new patterns which may be typical of other species. Other mutants are known which also induce the appearance of extra chaetes in different regions of the body surface. The new elements appear as discrete repetitions of already existing ones (*Tuft (Tft)*, *polychaetoid (pyd)*, etc.) or as an irregular increase of the number of elements, as for the sexcomb chaetes in *eyeless dominant (ey^D)*. It is questionable whether these addition pattern mutants actually determine the appearance of new patterns. We can interpret some of them as resulting from the uncovering of atavistic patterns, i.e. from the manifestation of preexisting developmental pathways.

Another group of mutants lead to a *substitution* of elements or entire regions of the body. We include here the homoeotic mutants (see GEHRING, this volume). The homoeotic transformation corresponds to few elements, to large regions of the body and even to complete segments. Dominant and recessive mutants have been isolated which change the segmental position of the sexcomb (*Polycomb (Pc)*, *extra sex combs (esc)*). Other homoeotic mutants transform rather large regions of the body, like the antenna (*Antennapedia (Antp)*, *aristapedia (ss^a)*) and proboscis (*proboscipedia (pb)*) into leg segments, eye facets into wing structures (*ophthalmoptera (ey^{opt}*) etc. Even larger transformations occur in mutants of the series "*bithorax*" (see LEWIS, 1968). Two recessive alleles *bithorax (bx)* and *postbithorax (pbx)* transform, respectively, the anterior and posterior part of both dorsal and ventral metathorax into mesothorax. Another recessive allele *bithoraxoid (bxd)* transforms the posterior part of the first abdominal segment into the posterior part of the metathorax, both dorsally and ventrally. Transformations in the opposite direction, that is from anterior to posterior segments, have also been isolated. Thus, *Contrabithorax (Cbx)* transforms meso into metathorax, *Contrabithoraxoid (Cbxd)* metathorax into the first abdominal and *Ultraabdominal (Uab)* the first abdominal into the second abdominal segment. Even within segments transformations between anterior and posterior structures have been observed in mutants. Thus *engrailed (en)* transforms the posterior regions of the thoracic segments into the anterior structures of the same segment. This transformation involves ventral structures as well as dorsal ones (GARCIA-BELLIDO and SANTAMARIA, 1972). The mode of action of these homoeotic mutants is as yet unknown. Homoeotic genes have been considered as genes whose normal function is that of repressing developmental pathways typical of homologous segments (LEWIS, 1964). What is interesting in this context is that homoeotic mutants do not create new patterns but determine the appearance of normal ones in other body regions.

Another type of substitution pattern mutants include those which interfere with the sex-determining genetic systems. Among the numerous intersexual or sex reversal mutants there are some which transform only part of the bisexual patterns. Thus, *sexcombless (sx)* suppresses the male sexcomb and also modifies the external genitalia, and *intersex (ix)* affects only females in the external genitalia. Again, these mutants do not lead to the formation of new patterns but to the shift of developmental pathways towards the formation of already preexisting patterns.

Some other mutants lead to a *disorganization* of normal patterns. Thus, eg., *dachs (d)*, *comb gap (cg)*, *humpy (hy)*, shorten, fuse or deform leg segments and entire body regions with frequent pleiotropic effects. These malformations could result from general alterations of growth processes in the imaginal disks. But a detailad developmental analysis could detect more specific mechanisms (see DATTA and MUKHERJEE, 1971).

This short presentation of pattern mutants reveals several constant features of cuticular patterns. Mutations may lead to the appearance of patterns of one region of the body in another (homoeotic mutants) or change morphogenesis and differentiation in the male or female direction (intersexual mutants). Other mutations result in the removal or addition of single and characteristic elements of a pattern, leaving the rest unchanged. Some of them merely shift development towards the formation of atavistic patterns. The addition or suppression pattern mutants indicate that specific genes may control the appearance of specific elements without interfering with the general pattern. However, the mutant patterns represent annoyingly few variations to the normal ones. Thus, apparently patterns result from the interaction of many discrete genetic factors.

C. Experimentally Induced Variations of Patterns

Experimental conditions may interfere with the normal development and lead to pattern variations. Chemical or physical agents applied at different developmental periods may induce morphological alterations of the adult pattern. Many of them correspond to unspecific malformations or to disorganized patterns. Some of them, however, involve alterations similar to the one found in pattern mutants. Thus, ether applied at early stages of the embryonic development produces, among others, phenocopies of mutants of the series *bithorax* (GLOOR, 1947). X-rays (WADDINGTON, 1942), mutagenic agents, such as nitrogen mustard (BODENSTEIN and ABDEL-MALEK, 1949), nucleic acid analogues, such as fluorouracil (GEHRING, 1964), applied during the larval development can induce pattern duplications and small homoeotic transformations. Heat shocks applied to pupae induce phenocopies in the final shape and venation pattern of the wing (HENKE, 1947; STUMPF, 1959). The mode of action of these agents is not known, but they possibly trigger the initiation of developmental pathways that are controlled by the normal genome.

As shown in transplantation experiments (NÖTHIGER, this volume) fragments of imaginal disks will differentiate only a fraction of the imaginal disk derivatives. When these fragments are cultured prior to metamorphosis, proliferative growth takes place. During metamorphosis, the grown fragment can differentiate in addition to its prospective structures, new ones which are of two main types: 1) On the one hand, they appear as *duplications* of the patterns typical for the original fragment (HADORN and BUCK, 1962; NÖTHIGER and SCHUBIGER, 1966; GEHRING, 1966). The patterns in such duplications are usually of normal size but sometimes restricted to parts only of the complete pattern inventory. These partial patterns could represent developmental units in pattern formation. In any case, their existence suggests independent developmental control of groups of structures or arrangements, within the normal large patterns; 2) On the other hand, fragments subcultured for several transfer generations may undergo *transdetermination* and give rise to cuticular arrangements typical of other imaginal disks (HADORN, 1966a; see GEHRING, this volume). The transdetermined structures are again, as a rule, arranged in normal patterns. The transdetermination experiments are of great relevance in the context of pattern formation. The new cuticular structures constitute normal and entire rather than abnormal or partial arrangements. However, incomplete patterns have occasionally been found, consisting of a few elements only. They probably correspond to incipient patterns. In some instances, disorganized structures of wing and leg chaetes (sex-combs) appear (HADORN, 1966a). Finally, in other cases subcultures were isolated, which recurrently differentiate abnormal patterns. Thus, an anormotypic line of anal plates differentiated anal-plate-like chaetes uniformly distributed over large areas, instead of being subdivided into definite regions as *in situ*. This and other abnormal patterns are thought to result from genetic mutations (HADORN, 1966b).

The preceding survey on pattern variations suggests that *patterns result from the mosaic contribution of independent "developmental pathways"* (LEWIS, 1963). Single patterns and even elements of a pattern can be experimentally or genetically affected independently of other elements or patterns. We do not know what are the material bases of such developmental pathways. If they are real entities, comparable to metabolic pathways, we should be able to describe them in terms of *genes* acting within *cells*. In the following chapters the underlying mechanisms of pattern formation will be discussed in these terms.

III. The Arising of Topological Differences

The characteristic distribution of the cuticular structures on the adult body surface is a final step in the development of the imaginal disks. Each adult cuticular structure represents the differentiation product, during metamorphosis, of a single imaginal disk cell. The cells of premetamorphic imaginal disks have an embryonic appearance throughout development (see URSPRUNG, this volume) with no morphological clues to anticipate their prospective differentiation. In principle, the distribution in patterns of the differentiated cells could result from last moment decisions imposed upon a population of undetermined and undifferentiated cells. However, indirect experimental evidence suggests that the differentiation of the cells in adult cuticular patterns results from stepwise decisions made throughout the previous development of the imaginal disks (see NÖTHIGER, this volume).

The continuous pattern of the adult integument results from a mosaic contribution of the different imaginal disks (ZALOKAR, 1943; MURPHY, 1967). The different imaginal disks of developing larvae are anatomically and morphologically identifiable (BODENSTEIN, 1950). These observations suggest that every single imaginal disk has a *particular developmental history*.

The analysis of gynandromorphs indicates that the epidermal cells of each imaginal disk derive from a discrete and small number of cells (STERN, 1940) present in a characteristic position on the early blastoderm cell layer (see GARCIA-BELLIDO and MERRIAM, 1969). The primitive ectodermal anlagen are set apart from the remaining larval epidermis, invaded by nervous connections and mesodermal derivatives, and start growing early in the larval period (AUERBACH, 1936). In most imaginal disks cell proliferation apparently occurs in a continuous and exponential way throughout the larval development (see NÖTHIGER, this volume). The role of neural and mesodermal elements in determining imaginal disk growth and morphogenesis is not known. However, as discussed below, transplantation and cell mixture experiments suggest that the cuticular characteristics of each imaginal disk are exclusively dependent on their epidermal components.

Several experimental approaches have revealed regional differences among the cells of a growing imaginal disk. These differences may be relevant to their prospective organization into patterns. In gynandromorphs male-female boundaries may run between any given pair of structures (STURTEVANT, 1929). This indicates that no fixed cell lineages are established prior to blastoderm formation. The frequency with which two structures are of different sex can then be used to construct morphogenetic maps of the location of the presumptive adult structures in the primitive anlage of a given disk. A comparison of the map obtained in this way with the corresponding map of the adult structures permits the detection of morphogenetic particularities in the development of that imaginal disk. Thus, RIPOLL (1972) studied the growth of the dorsal mesothoracic disk anlage (Fig. 3). Whereas the primitive anlage can be mapped in a surface, the imaginal disk seems to grow into a cone, in which the broad tip corresponds to the future wing margin, the dorsal sides of the cone will give rise to the mesonotum and dorsal surface of the wing, and the ventral side to the ventral surface of the wing and pleurae. The same description probably applies to the development of other adult appendages. The protrusion of the cone tip out of the original plate could be achieved by several morphogenetic mechanisms: 1) by local mechanical stretching, 2) by directed migration of cells or cell groups, 3) by differences in the growth rate between the centre and the periphery of the plate, 4) by intercalar cell death in the periphery of the plate (SPREIJ, 1971), 5) by preferential orientations of the mitotic spindles in successive mitoses.

Genetically marked clones resulting from induced mitotic recombination (see NÖTHIGER, this volume) are elongated in the proximo-distal axis in the wing (GARCIA-BELLIDO, 1968b; GARCIA-BELLIDO and MERRIAM, 1971a; BRYANT, 1970) in the leg (TOKUNAGA, 1962; BRYANT and SCHNEIDERMAN, 1969) and in the antenna (POSTLETHWAIT and SCHNEIDERMAN, 1971), that is in body appendages. Interestingly, in the mesonotum clones are more or less isodiametric (BRYANT, 1970; GARCIA-BELLIDO and MERRIAM, 1971a; MURPHY and TOKUNAGA, 1970). A more detailed analysis of the clone profiles in the wing surface shows that these are irregular, with indentations perpendicular to the main orientation of the clone (Fig. 5a). This renders improbable the alternative of mechanical stretching. Moreover, the clones appear in compact units and are only rarely split in small fractions, which suggests that cell migration is rare. The sizes of mosaic patches, in terms of cells, are, with small variations, uniform in the surface of the adult appendages. This is so at least

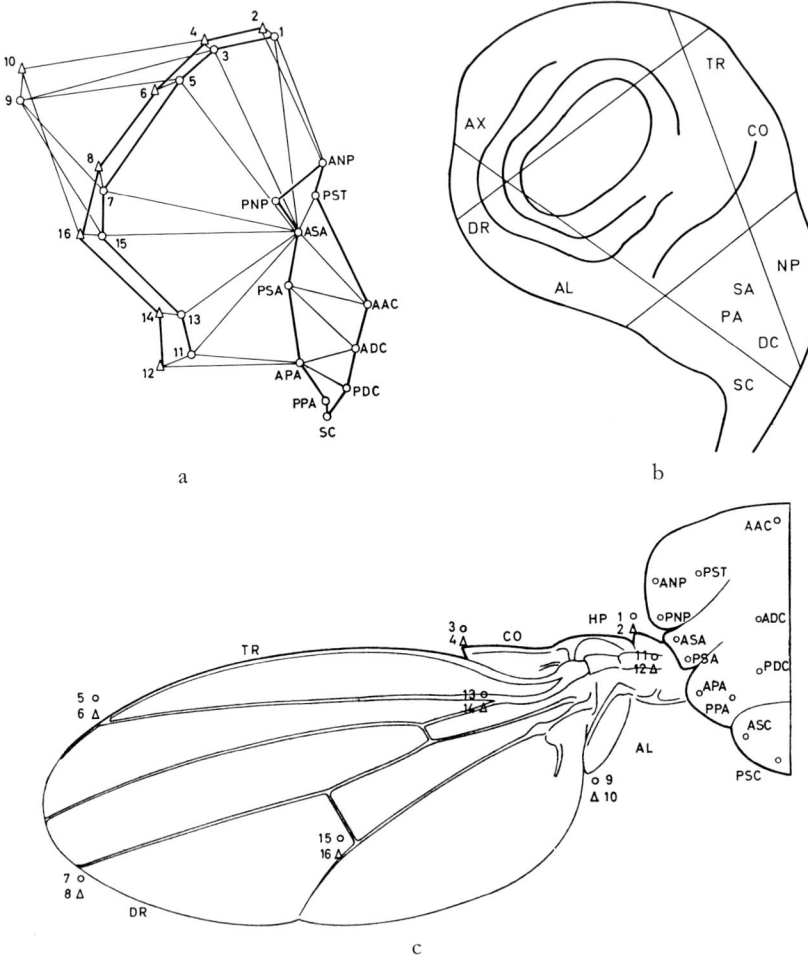

Fig. 3. Different stages in the development of the dorsal mesothoracic disk. a Anlage plan of the location of the presumptive adult structures in the blastoderm (RIPOLL, 1972); b Morphogenetic map of the imaginal disk in mature larvae (HADORN and BUCK, 1962); c Adult map (same labelling as in a). In b *Ax* corresponds to the wing hinge. *NP* notopleural; *SA* supraalar; *PA* postalar; *DC* dorsocentral; *SC* scutellar pairs of chaetes. (*A* anterior and *P* posterior)

in the dorsal mesonotum and wing blades of the wing disk (GARCIA-BELLIDO and MERRIAM, 1971a) indicating, that if cell death occurs, it is not in these prospective regions of the imaginal disk. BRYANT (1970) found lower clone sizes in the mesonotum than in the wing, but this possibly derives from the fact that he measured planimetric surfaces, and did not correct for differences in cell densities. Thus, cell proliferation is possibly similar in all regions of the wing disk throughout development, which makes alternatives 3 and 4 unlikely. The profile of the clones in both mesonotum and wing suggests that the *orientation of the mitotic spindles* plays a fundamental role in creating the adult form. Changing orientations of successive mitoses were histologically detected in the pupal wing and found to be region and stage specific (STUMPF, 1956). The role of oriented mitoses in morphogenesis has been emphasized by POSTLETHWAIT and SCHNEIDERMAN (1971) in a comparative analysis of the shape of clones in the cuticle of different insects.

Analysis of the type of structures found in clones after mitotic recombination uncovers other regional characteristics in the development of the imaginal disks. Whereas the analysis of gynandromorphs suggests that no strict regional cell lineages exist in the primitive anlage, *cell lineage restrictions* appear during later development. Patches in gynandromorphs run, as a rule, parallel to the veins but frequently cross over them and cover neighbouring intervein regions. The only exception is vein 4 in both dorsal and ventral surfaces. In gynandromorphs, mosaic borders are much more frequent in the intervein region 3—4, close to vein 4, than in other veins, intervein regions and wing margins (RIPOLL, 1972). This is especially interesting since vein 4 marks the separation between anterior and posterior in the homoeotic transformation of *bithorax, postbithorax* and *engrailed* (Fig. 5). This morphogenetic restriction is the earliest indication of separate cell lineages. It is followed later in development by others. Whereas mitotic recombination induced in the embryo and early first instar leads to clones embracing structures of the mesonotum, dorsal and ventral surfaces of the wing, as in gynandromorph patches, recombinant clones initiated in the late first instar do not include structures of both mesonotum and wing, or both dorsal and ventral surfaces of the wing (BRYANT, 1970). Especially the last finding is meaningful since clones may run along the wing margin for hundreds of cells without crossing the dorso-ventral separation line (GARCIA-BELLIDO, 1968b; GARCIA-BELLIDO and MERRIAM, 1971a; BRYANT, 1970). Neither do clones cross over vein 4 of the wing (see (Figs. 3, 4 and 5 in BRYANT, 1970). Some clonal restrictions seem to separate anterior and posterior, dorsal and ventral wing surfaces and possibly wing and mesonotum regions at a stage in which the mesothoracic disk contains no more than 50—100 cells. The orientation of the clones is again relevant here. As seen before clones in the wing surface are elongated, whereas they are more or less isodiametric in the mesonotum. Within the wing surface, clones have different main orientations in the anterior and posterior halves. Clones in the anterior part run parallel to the margin, while they meet the margin at different angles in the posterior part (Fig. 5a). Also the mitotic rate seems to be different along the wing disk surface. At a given age of larval development clones are smaller in the anterior part of the wing and notum than in the posterior part. This difference disappears in imaginal disks of mature larvae (GARCIA-BELLIDO and MERRIAM, 1971a).

Differences in the prospective differentiation of the growing cells have been detected in the mesonotum. MURPHY and TOKUNAGA (1970) have shown that in clones initiated at different ages of development, some adult chaetes of the mesonotum are more frequently within the same clone than others located at similar distances. The authors suggest that these topological correlations could indicate common cell lineages. However, since these correlations are only statistical they probably reflect preferential cell orientations rather than strict cell lineages (see RIPOLL, 1972). Similar cell lineage restrictions have been found in other imaginal disks (see NÖTHIGER, this volume). The pioneering work of BECKER (1957), on the clonal analysis of eye development, shows that in the late first instar when the eye disk contains about 150 cells, specific spindle orientations start separating clones of the dorsal and ventral halves of the eye anlage. Later in development, probably during the third instar, presumptive cells for head cuticle and ommatidia become separated by cell lineage. At this time clones become also restricted to single segments of the leg (BRYANT and SCHNEIDERMAN, 1969) and the antenna (POSTLETHWAIT and SCHNEIDERMAN, 1971).

Fragmentation experiments indicate that in imaginal disks of late third instar (mature) larvae, cells or groups of cells have very restricted potentialities (see NÖTHIGER, this volume). Cell lineage studies corroborate this conclusion. In the wing disk clones initiated 40 h before pupariation no longer embrace simultaneously chaetes and trichomes although they may contain several of these elements alone (GARCIA-BELLIDO and MERRIAM, 1971a). Eight hours before pupariation the cells giving rise to either chaetes or trichomes are already irreversibly committed (GARCIA-BELLIDO and MERRIAM, 1971b). At this time, divisions in the disk are coming to an end. The differential divisions of the chaete-forming cells start at about pupariation, as indicated by histological criteria (LEES and WADDINGTON, 1942), their high sensitivity to phenocopying agents, and the drop in responsiveness to induced mitotic recombination (GARCIA-BELLIDO and MERRIAM, 1971a). Mitoses in the wing disk, and possibly in other imaginal disks, cease at about 18—21 h after puparium formation (STUMPF, 1956; GARCIA-BELLIDO and MERRIAM, 1971a). At this stage differentiation of the epidermal cells into cuticular structures and therefore into patterns sets in.

As we have seen, the development of an imaginal disk proceeds stepwise, delimiting cell lineages characterized both by their regional position and by their prospective differentiation. Thus, *pattern formation apparently does not occur by a sudden differentiation of still totipotent cells over the entire mature disk*. If during normal development regional differences within a disk arise, and these correspond to established cell lineages, localized cell differentiation could be a direct consequence. The following analysis will show how far this inference stands the experiments aimed to test it.

IV. The Final Pattern

As shown in the foregoing section the imaginal disks, along with the increase in cell number, acquire more and more specialized regional or topological differences. The imaginal disks of mature larvae consist of a mosaic of cells, or groups of cells, with the presumptive characteristics of the adult pattern. From fragmentation experiments an anlage plan of the location of the presumptive cuticular structures in a given imaginal disk can be constructed (see NÖTHIGER, this volume). However, the detailed organization varies among different imaginal disks. In the male foreleg imaginal disk of mature larvae the location of the presumptive cells for the major cuticular structures of the adult pattern can be pinpointed in the anlage plan (Fig. 1a) (SCHUBIGER, 1968). In the female (HADORN and GLOOR, 1946) and in the male genital disk (HADORN et al., 1949; URSPRUNG, 1959; LÜÖND, 1961), the same experimental approach can only resolve regions with presumptive characteristics for complete patterns (Fig. 2). The imaginal disks of the tergites of mature larvae cannot be subdivided into presumptive regions of the adult pattern (GARCIA-BELLIDO and MERRIAM, 1971c); SANTAMARIA and GARCIA-BELLIDO, 1972). It is possible that these differences among disks depend on the developmental stage of each disk. These stages could differ for different disks if we express them in terms of the number of cell divisions occurring until differentiation, rather than in terms of hours of larval development.

These observations suggest that the determination and arrangement of single presumptive cells into patterns takes place in the last stages of development. Several experimental approaches have been used to find out at which moment and by which mechanisms single cells become determined to differentiate into a given cuticular structure in a given location. There are two alternative mechanisms of pattern formation, which can be now stated in more specific terms: 1) on the one hand, subsequent divisions will segregate during development progressively more specific prospective differentiations in such a way that the mature disks already consist of mosaically distributed presumptive cells for every single cuticular structure. Thus, the adult pattern would result as a developmental consequence of segregations of cell lineages for topological and structural characteristics. 2) On the other hand, proliferating cells would segregate into groups or blastemas with properties of a morphogenetic field in which the single undetermined cells will differentiate according to their position in the system.

These alternatives can be tested experimentally: How do cells respond to the influence of other cells derived from foreign blastemas or carrying different genetic information?

A. Cell Lineage

Genetic mosaics constitute experiments in which cells of different genotypes are confronted with each other in very intimate "cell mixtures". Since mosaicism can be induced at a given time, any possible interaction between different cells can be referred to specific moments of development.

Male and female *Drosophila* flies differ in some of their adult cuticular patterns (Fig. 1b, c). In gynandromorphs, when the male-female demarcation line runs through a sexual dimorphic pattern, such as the basitarsus of the foreleg, the male and female cells differentiate their own structures *autonomously* (STERN and HANNAH, 1950). A foreleg basitarsus can be mosaic for chaetes of the transverse rows, which are similar in both sexes, but only the male cells develop into sexcomb teeth. A single tooth can show autonomous differentiation even when surrounded by female cells. Conversely, a single female chaete can be found surrounded by sexcomb teeth. Moreover, the *position* of the sexcomb teeth in the mosaic basitarsus corresponds approximately to the position of these cells in an entirely male basitarsus. Cases of abnormal position of the sexcomb teeth or of the female chaetes were interpreted as being due to the influence of the neighbouring cells of the opposite sex. And yet, the most consistent finding was that the resulting adult pattern appeared as a hybrid and integrated pattern, as if the male and female patterns, although different, actually shared the same building principle or the same "prepattern" (STERN, 1954). The differential response of male or female cells would then merely reflect the autonomy of the expression of their genetic constitutions. Similar analyses of the external genitalia of gynandromorphs showed autonomous differentiation of the male and female cells in mosaics as well (KROEGER, 1959). However, in this case the male and female structures did not appear intermingled, but rather were segregated into male and female patterns. Here, the common prepattern can only be referred to the general organization of the disk, rather than to the fixed and definitive location of every single pattern element.

Other experiments were carried out in individuals mosaic for normal and pattern mutant tissues. One of such mutants is *achaete (ac)*, a recessive pattern suppression mutant (see p. 64). STERN (1954) tried to analyse the mechanisms determining the suppression or appearance of a certain chaete of the notum, normally absent in *ac* flies. In mosaic flies of the genotype $y^+ac^+/y\ ac$, recombinant homozygous *ac* clones were marked by being simultaneously *yellow (y)*, a cell marker mutant which changes the normal dark pigmentation of the chaete into light yellow. In almost all cases, and independently of the amount of ac^+ or *ac* tissue surrounding the location of this chaete, it was present when the tissue occupying the chaete location was ac^+, and absent when it was *y ac* (Fig. 4a). These results indicate that *ac* does not determine a cuticular pattern differing from that of ac^+. If this were the case, the factor determining the appearance of a chaete or its absence should depend on the nature of the prevailing tissue around the chaete location. Similar to the conclusion gained from the bisexual patterns of gynandromorphs, it seems that *ac* does not change the normal pattern, but merely renders the cells in a given location unable to respond positively to the signal arising from the unchanged prepattern. Thus, *prepatterns could represent localized inhomogeneities in the growing cell population imposing upon the cells very specific developmental alternatives*. The signal to produce a chaete is, however, not absolutely restricted spatially, since a normal y^+ac^+ chaete may appear slightly displaced from the normal location, if this is occupied by *ac* tissue. These small inexactitudes in mosaics suggest that these prepattern inhomogeneities, as in morphogenetic fields, are distributed over a range of several cells, or lead to the production of diffusible substances to which single undetermined cells can respond (STERN, 1954).

Similar experiments were extended to other mutants determining a change in the final pattern, with identical results. Cells behave autonomously according to their own genetic constitution differentiating or not a given cell structure in an integrated pattern with normal cells (see STERN, 1965 for review). This situation was also true in mosaics for homoeotic mutants such as *aristapedia (ss^a)* (Fig. 4b) (ROBERTS, 1964). In $y\ ss^a/y^+ss^{a+}$ flies the recombinant $y\ ss^a$ spots invariably differentiate tarsal structures if they occur in the distal part of the antenna. In this case, a consequent interpretation of the results would mean that adult patterns as different as antenna and leg have a similar prepattern to which normal and ss^a cells will respond differentially depending on their genotype. This, in principle, is not surprising since antennae and legs are possibly homologous organs. However, if the prepattern is the same, why then do normal cells respond differentially in the leg and in the antenna? One would have to assume that not only cells differing in genotype but also in location or developmental history respond differentially to a given constant prepattern. This conclusion weakens the operational value of the prepattern concept, at least for the pattern mutants studied so far.

Although prepattern mutations leading to early developmental changes would presumably be lethal, it is reasonable to assume that mutations, leading to late or small prepattern changes would be detectable. Possibly one such mutant is *eyeless dominant (ey^D)*, which determines the appearance of extra sexcomb teeth in an enlarged male basitarsus (STERN and TOKUNAGA, 1967). By mitotic recombination it

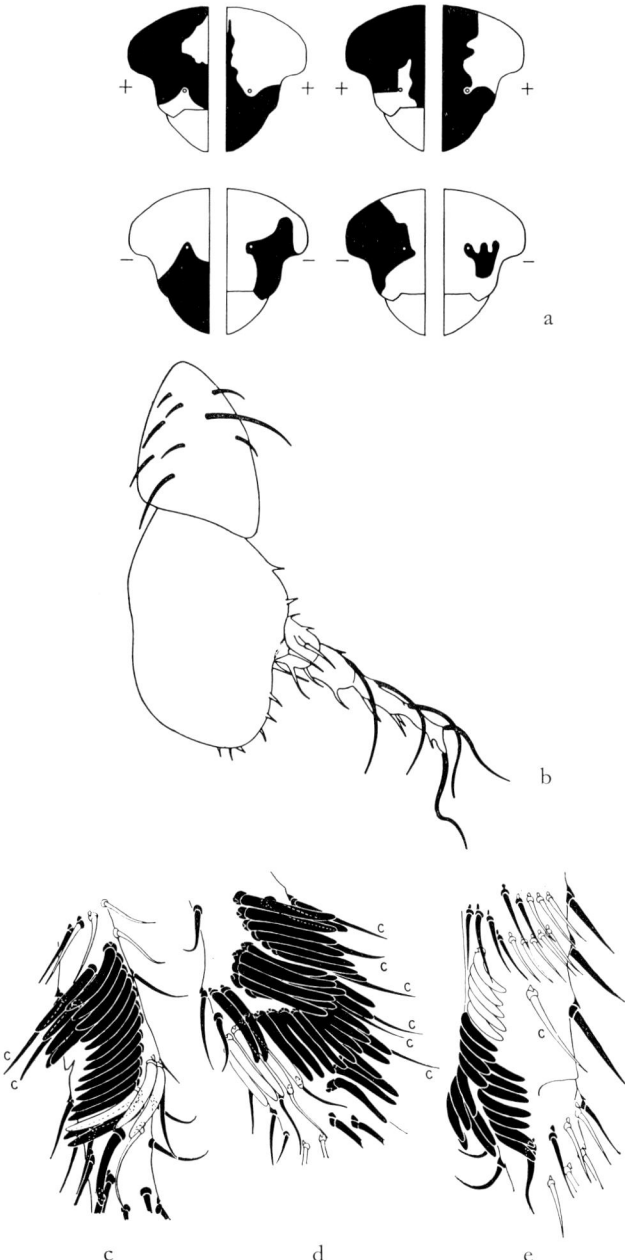

Fig. 4a—e. Mosaic patterns resulting from mitotic recombination during development. a mosaicism of *achaete* (*ac/ac* black) and normal (*ac/ac*$^+$, white) territories determining the autonomous presence (+) or absence (−) of a given chaete in the notum (STERN, 1954;) b mosaicism of *aristapedia* (*ss^a/ss^a*, white) and normal (*ss^a/ss^{a+}*, black) territories and the autonomous manifestation of leg or antennal structures (ROBERTS, 1964); c—e mosaicism for *ey^D*/+ (black), *y* and presumably *ey^{D+}* (white) structures. c, d possible non-autonomy, e possible autonomy of presumable *y*; *ey^{D+}* cells (STERN and TOKUNAGA, 1967)

should be possible to mark clones of normal cells in an heterozygous $y^+ ey^D/y ey^{D+}$ background. Unfortunately, the cell marker mutant used, *yellow*, lies distal to the location of ey^D, and thus y clones do not necessarily mark the ey^{D+}/ey^{D+} constitution. Several clones of y cells differentiating extra sexcomb teeth were found. If they were simultaneously ey^{D+} their appearance would suggest their non-autonomous development in the presence of surrounding ey^D tissue (Fig. 4c—e). This non-autonomy could indicate the existence of a mutant prepattern determining mutant pattern differentiation of genetically normal cells.

Hairy wing (Hw) leads to a higher than normal chaete density in certain regions of the mesonotum. This mutant has been studied in clones following mitotic recombination, and apparently Hw^+ tissue still retains the chaete density of the neighbouring $Hw/+$ territory (GOTTLIEB, 1964). Thus, it could represent another prepattern mutant. *Hairy wing* also determines the appearance of extra chaetes on some veins of the wing surface, normally devoid of them (GARCIA-BELLIDO and MERRIAM, 1968, 1971b). When mitotic recombination in $Hw/+$ cells produces Hw^+/Hw^+ clones, detectable by closely linked genetic cell markers, they do not produce extra chaetes in the wing surface although surrounded by Hw tissue. Yet, when the mitotic recombination occurred later than 8 h before pupariation Hw^+/Hw^+ cells still differentiated into extra chaetes. Hw/Hw clones resulting from the same recombinational event always differentiate extra chaetes. The existence of a recessive pattern addition mutant, such as *hairy (h)* which mimics the Hw phenotype, permits us to perform the reciprocal experiment. Clones of h/h cells in a $h/+$ background are able to differentiate extra chaetes when they are initiated during the larval development up to 8 h before pupariation. The recombinational event leading to the production of h/h clones is not manifested in the appearance of extra chaetes if it occurs later. In the Hw and in the h experiments the results are similar. Whereas a change in the genetic constitution earlier than 8 h before pupariation is accompanied by a corresponding change in the phenotype, the same change after that moment remains without effect (GARCIA-BELLIDO and MERRIAM, 1968, 1971b). Possibly, the information needed for cell determination is already processed in the mother cell and this information persists after a change in the genetic constitution in the daughter cells. In principle, this "perdurance" could explain some cases of cell non-autonomy in genetic mosaics (GARCIA-BELLIDO and SANTAMARIA, 1972).

The Hw and h extra chaetes appear only in certain clones of the mesonotum, scutellum and certain wing veins. Only clones of mutant cells in these regions will produce extra chaetes. We do not know the nature of the localized singularities which determine a common prepattern for Hw, h and normal individuals, nor how the prepattern information is transmitted to the cells. However, an alternative hypothesis could explain these results. Only cells located by cell lineage in the appropriate regions, are competent to respond to an unspecific signal; the nature of their response would depend on the cell genotype. The emphasis of this interpretation lies on the competence of the cells rather than on the local, specific signals.

Localized inhomogeneities determining a prepattern become more difficult to explain the younger the imaginal disk is and the fewer cells it contains. A clonal analysis of homoeotic mutants such as *Antennapedia* (POSTLETHWAIT and SCHNEIDERMAN, 1969) and *engrailed* (GARCIA-BELLIDO and SANTAMARIA, 1972) (Fig. 5) shows that normal and mutant growth patterns are different from early stages of development. Here a postulated common prepattern between homologous organs would have to work in spite of, or preceding, the different growth patterns.

The experiments with genetic mosaics show that, as a rule, *cells differentiate autonomously according to their genetic constitution*, even when in intimate cell contact with cells of other genotypes. The actual location of normal and mutant structures

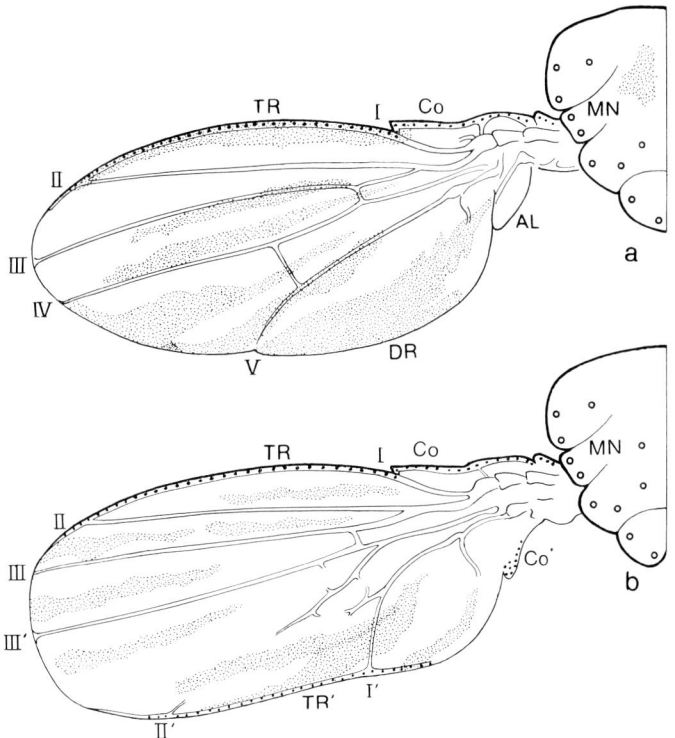

Fig. 5a, b. Comparison of clones of trichomes in the wing of normal (a) and *engrailed* (b) flies. *Mn* mesonotum; *Co* costa; *TR* triple row of chaetes; *DR* double row; *AL* alula. *I–V* wing veins. Labelled with (') the homoeotic structures. ○ and ●, chaetes (GARCIA-BELLIDO and SANTAMARIA, 1972)

in an integrated pattern may result from the differential response of undetermined cells to the same underlying prepattern as well as from other morphogenetic mechanisms. We have seen that in gynandromorph patches the teeth in the sexcomb area do not always appear in the correct position, and that the male structures of the genitalia do not intermingle with the female structures. Cell lineage studies indicate that the presumptive sexcomb cells become shifted relative to the remaining cells, to reach their final position (Fig. 7a) (TOKUNAGA, 1962). A segregation of sexcomb teeth and female chaetes has been found in aggregation experiments of male and female foreleg cells (GARCIA-BELLIDO, 1966a). NÖTHIGER (1964) described a similar segregation of male and female cells corresponding to some adult patterns in aggregates of male and female genital disk cells. In cell aggregates of *aristapedia* antennal cells and normal leg cells we have only found mosaicism in leg structures, while aggregates of normal leg and normal antenna cells result in an almost complete segregation of both kinds of cells (GARCIA-BELLIDO, 1968a). Thus, these segregations of male and female cells and of normal and mutant cells corresponding to different presumptive patterns are indicative of cell differences which precede the location of these cells in patterns (see below).

B. Cell Affinities

As seen before, fragmentation experiments reveal a mosaic organization of presumptive patterns and cuticular structures in mature imaginal disks. However, these experiments do not tell us whether the information to organize the adult pattern lies in the single cells themselves or in groups of cells with characteristics of a prepattern or morphogenetic field. To test the specific information of single cells we should be able to study the differentiation of isolated cells. Since, so far, this is technically not possible, an alternative experimental approach is currently used. Disks can be readily dissociated into single cells and subsequently reaggregated into pellets which can then be handled like imaginal disks in transplantation experiments (see NÖTHIGER, this volume). If the imaginal disks are genetically labelled with cell marker mutants, the adult structures can be distinguished as to their origin by the mutant phenotypes. HADORN et al. (1959) first showed that cell aggregates derived from different, genetically marked wing imaginal disks were able to differentiate adult wing cuticular structures arranged in patterns like those found *in situ*. The integrated normal patterns could result from the response of undetermined cells to a morphogenetic field organization which would spontaneously arise in the mixed cell population (URSPRUNG and HADORN, 1962). If this is the correct interpretation, a mixture of cells from different imaginal disks might lead to an "assimilative induction" of cells of a disk under the influence of the cellular reorganization of the other disk. NÖTHIGER (1964) performed such an experiment combining cells of the genital and the wing disk. He found both, *separation of the two kinds of mixed cells and gathering of cells of each type, differentiating patterns typical of their disk of origin.*

Similar experiments combining cells of leg and wing disks of mature larvae (GARCIA-BELLIDO, 1966b, TOBLER, 1966) antenna and wing (GEHRING, 1966), antenna and leg (GARCIA-BELLIDO, 1968a) resulted in the separation of territories of each kind of cells organized in patterns similar to those found *in situ*.

The differentiation side by side of cells derived from different imaginal disks (heteronomous combinations) and the differentiation of integrated mosaics in aggregates from identical disk cells (homonomous combinations) can result from several mechanisms. The trypsin treatment used to dissociate the imaginal disks produces a cell suspension consisting of more than 90% single cells and clusters of few (2—20) cells. A vital staining test indicates a cell mortality not surpassing 2% of the cells (GARCIA-BELLIDO, 1966a, 1967). However, it could be argued that the reaggregates only consist of descendents of these groups of cells, because single cells would be irreversibly damaged. Aggregates of trypsin dissociated cells of imaginal disks cultured for 2 days *in vivo* in starved adult hosts, which prevents proliferative growth, appear as histologically organized implants of the size of the implanted pellet. In these implants, the histotypical characteristics of leg reaggregates are very different from those of wing imaginal disk reaggregates (GARCIA-BELLIDO, 1967). These results indicate that single cells are able of reaggregating into morphogenetic systems: 1) with histotypical organizations characteristic of the tissue they derive from, 2) without cell proliferation, and 3) prior to the structural differentiation taking place during metamorphosis. Similar observations were obtained with embryonic cells (LESSEPS, 1965) and with other imaginal disk dissociates (KURODA, 1969) cultured *in vitro*. Thus, the association of cells into integrated patterns in homonomous combinations, and the separation of cells in heteronomous combinations possibly result from a mutual *sorting-out* of unlike (heterotypic) cells, so that, in the end, like (isotypic) cells come to lie together. Death of single cells would still allow mosaicism of the large territories. And yet, single cells or groups of few cells were actually found as mosaic patches in homonomous combinates or trapped within foreign territories ("faulty mosaics") in heteronomous combinates. If sorting-out is the basis of this cellular behavior, we have to assume that the cells have *recognition properties* which distinguish isotypic from heterotypic cells (TOWNES and HOLTFRETER, 1955).

Recognition specificities of mature imaginal disk cells go beyond simply telling apart cells from different imaginal disks. Marked cells deriving from anterior and posterior, or proximal and distal fragments of the wing disk (GARCIA-BELLIDO, 1966b) and cells from proximal and distal fragments of the male foreleg (TOBLER, 1966) segregate, building separate patterns. In combinates of cells deriving from entire imaginal wing disks and cells of fragments of the anterior part mosaicism appears only in anterior patterns (GARCIA-BELLIDO, 1966a). The different pairs of legs have common patterns except for some characteristic regions (Fig. 1b, e). The basitarsus of the first leg pair contains transverse rows of identical chaetes in the anterior part of the leg. These transverse rows are not present in the second pair of legs but appear again with identical distribution in the third leg, although here they are located in the posterior part. Combinates of imaginal cells of the different pairs of leg disks show mosaicism in all the common leg patterns. However, we did not find mosaicism in the transverse rows in first and second, nor even in first and third leg cell combinates. Male and female cells of the first leg basitarsus appear separated in the sexcomb area, although mosaically integrated in transverse rows, a pattern common to both sexes (GARCIA-BELLIDO, 1966a). Male and female cells of the genital disk form mosaic structures in the anal plates but not in other bisexual patterns (NÖTHIGER, 1964). Conversely, different cuticular structures corresponding to homologous patterns of *D. melanogaster* and *D. séguyi* can appear in integrated patterns in cell combinates of male genital disks of these species (NÖTHIGER, 1964) (Fig. 6a, b). Similarly, arista and claw, tarsal and antennal cells, also appear side by side ("heteronomous associations") in combinates of leg and antenna cells (GARCIA-BELLIDO, 1968a). These results indicate that *cell recognition reflects cell characteristics which are not directly related to the prospective structural cell differentiation*. Cells of identical prospective differentiation, such as the transverse row chaetes, may not be able to integrate in mosaics if they do not derive from the same patterns and, conversely, cells differentiating distinct cuticular structures, such as the different elements of a given pattern, may integrate in mosaics. Finally, associations between cells of homologous or similar patterns can occur although each cell type differentiates according to its origin. Cell recognition possibly reflects characteristics of actual cell-membrane-differentiation. The previous analysis suggests that their characteristics are not related to any specific cuticular differentiation, but rather to the position of this cell within a system. If the cell recognition characteristics are of a positional nature they should be found in cells in any stage of the imaginal disk development.

Specificity in cell recognition explains the sorting-out of cells derived from different disks, disk regions and patterns, and possibly also explains the *reconstruction*

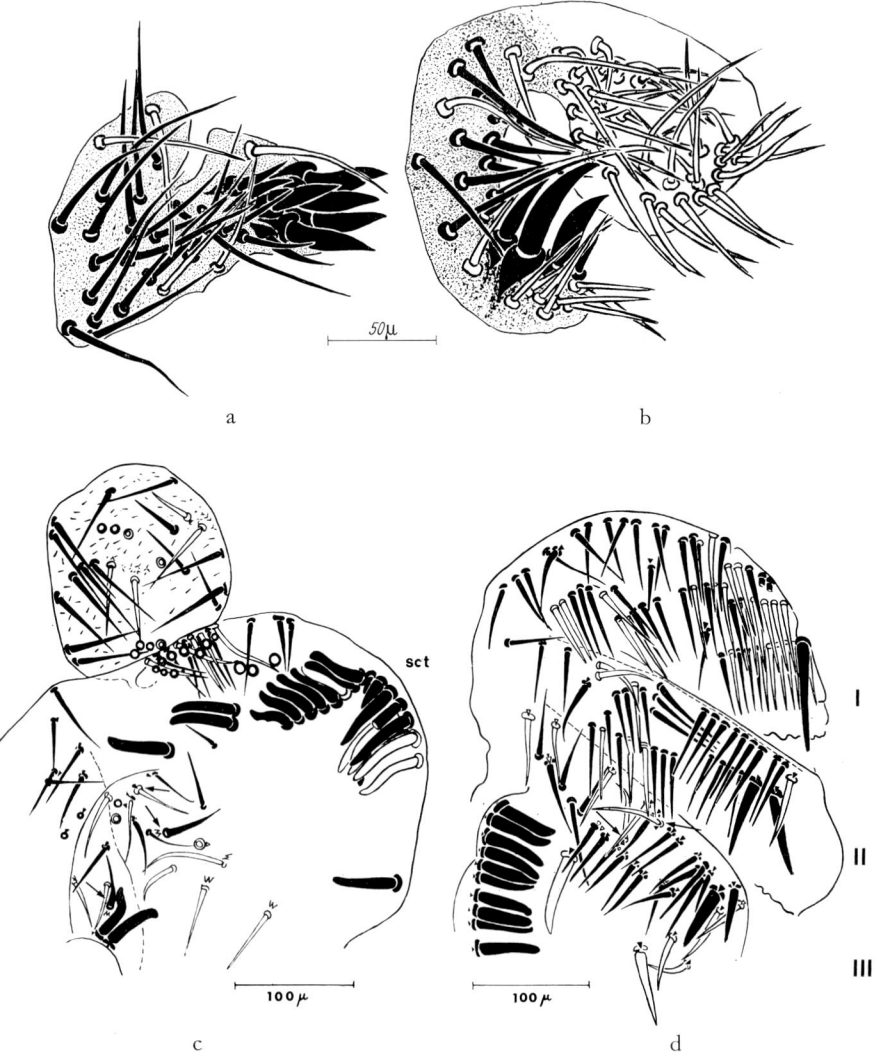

Fig. 6. Mosaic patterns after cell dissociation and reaggregation. a, b mosaicism in male genital structures (anal plates) of cell mixtures deriving from *D. séguyi* (black) and *D. melanogaster* (white) imaginal disks (Nöthiger, 1964); c mosaicism in basitarsal structures. Observe the arrangement of the sexcomb teeth (sct), isolated, grouped, aligned. Arrow: chaete-bract mosaicism; d mosaicism in tibial (I, II) and basitarsal (III) patterns of cells deriving from male foreleg disks. White and black structures: different marker mutant phenotypes. Arrows indicate "faulty orientation" of some chaetes and their bracts with respect to the remaining pattern (Garcia-Bellido, 1966a)

of patterns in mosaics. In aggregates of mature imaginal disk cells integrated patterns correspond, as a rule, to small body regions. The reconstructed patterns are *similar* to those found *in situ*. However, they appear frequently incomplete, or by far exceeding the number of normal elements, and arranged in several degrees of "correctness", compared to those found *in situ* (GARCIA-BELLIDO, 1966a). When we look for elements of linear repetitive patterns, such as the transverse rows and the sexcombs of the legs, they frequently appear in reaggregates as irregular gatherings and only occasionally mosaically arranged in rows (Fig. 6c, d). It is difficult to ascertain whether these mosaically reconstructed patterns result from the reassembly of independently determined cells, or whether, alternatively, a new pattern is imposed on undetermined cells.

In the experiments mentioned above single imaginal disk cells seem to differentiate autonomously, irrespectively of their position in the reaggregate. NÖTHIGER (1964) described cases of single identifiable structures of the genital apparatus trapped within wing surfaces. These "faulty mosaics" have been found in further experiments with leg and wing disk cells as cell structures differentiating misplaced in foreign territories (GARCIA-BELLIDO, 1966a). The actual differentiation of these structures could be casual, but more likely reflects the state of determination these cells actually had before dissociation. As seen before (p. 70) prospective chaeteforming cells are already determined, as e.g., in the wing disk of mature larvae. If single cells are specifically determined to differentiate a given cell structure at the time of dissociation, the different degrees of correctness of reconstructed patterns could possibly reflect stages in the process of rebuilding a pattern. Incorrect patterns, which are the most frequent, do not suggest a new formation of patterns by undetermined or dedetermined cells. Moreover, in reconstructed patterns "faultily oriented" elements are occasionally found, e.g. chaetes aligned in a mosaically reconstructed row, with a different orientation relative to the remaining chaetes. If they carry a "bract" the latter positively marks the wrong orientation (Fig. 6d). This finding suggests that the chaete and the bract behave as a unit in the reconstruction of the row. Faulty orientations indicate polar singularities in imaginal disk cells prior to dissociation. At the same time they make it easier to understand how pattern reconstruction could occur.

We imagine *pattern reconstruction in aggregates* of imaginal disk cells as resulting from the specific affinities or recognition properties of cells carrying the characteristics of their previous position in the pattern. As seen before cells do not gather because they have a similar prospective structural differentiation, but because they belong to the same region. They do not necessarily meet together because of directed migration, but possibly because they remain together following random collisions in a segregating cell population. This explains the fact that reconstructed patterns correspond to small cuticular surfaces. They arrange in rows or in two-dimensional patterns because they have polar singularities matching with similar elements in a row as well as with different elements outside of the row. We are aware that such specifities far exceed those found in cell aggregation experiments in other organisms (see CURTIS, 1967). An explanation of pattern reconstruction in terms of quantitative cell adhesiveness (STEINBERG, 1970) would require too many degrees of differential cell adhesiveness, or too long sorting-out periods to achieve it. And yet, cell aggregates implanted into mature larvae complete pattern reconstruction before metamorphosis, a time corresponding to 24—48 h. In any case, the cell recognition specificities theoretically required to explain pattern reconstruction do not exceed the recognition specificities required to explain cell segregations, which are a directly observed fact. The required specificities for cell recognition could be provided by specific genetic signals. We have shown that transdetermined cells (GARCIA-BELLIDO, 1966b) as well as mutant homoeotic cells (GARCIA-BELLIDO, 1968a) possess affinities corresponding to the new prospective differentiation. However, this does not necessarily indicate that specific genes control cell affinity since the tested cells derived from already formed blastemas and therefore presumably carried also the new positional information.

While the previous interpretation seems to apply to aggregates cultured for short periods of time before cell differentiation, other mechanisms may be involved in pattern reconstruction of aggregates kept in culture for longer periods. It has been observed that reconstructed patterns are more complete in numbers and types of elements the longer the aggregates were cultured prior to metamorphosis (GARCIA-BELLIDO, 1966b). Prolonged culture is associated with cell proliferation and with larger patches of non-mosaic territories. Thus, the more complete patterns may result either because the newly produced cells acquired the determination qualities of the lacking elements (regeneration), or because new patterns were formed *de novo* from old and new cells. The latter mechanism is apparently the one suggested by POODRY et al. (1971) under the operationally ambiguous term of "repatterning". However, since the new cells did not differ in genotype from the old ones the data at hand cannot distinguish between the two alternatives. If these complete patterns depended on cell proliferation for their organization we must postpone their interpretation until we will have a better understanding of the mechanisms of cell determination in regenerates (see p. 84, and NÖTHIGER, and GEHRING, this volume).

Differential affinities have also been studied with cells from several stages of development. Embryos of 10 h cut into anterior fragments, dissociated and combined with cells of entire embryos of a different cell marker genotype, led, after culture in the adult and subsequent implantation into metamorphosing hosts, to implants with mosaic territories (SCHUBIGER et al., 1969). The mosaicism was restricted to patterns of cephalic imaginal disks when the combination contained cells from anterior embryo fragments. Mosaicism in abdominal imaginal disk derivatives was found in a complementary experiment with posterior embryo fragments. CHAN and GEHRING (1971) extended these experiments to blastoderm embryos with similar results. The frequency of mosaicism was certainly low in both cases, indicating that the adult cuticular structures originated from a small number of cells. If mosaicism was the result of cell recognition by the primitive imaginal disk cells it would reflect differences among cells in the early embryo. However, in these experiments the observed mosaicism could result from cell affinities of already grown blastemas once their cells had completed their larval development in the adult milieu. This is what actually seems to happen in transdetermined cells. Whereas leg and wing cells first segregate from each other, subsequent spontaneous transdetermination of leg into wing cells can eventually give rise to wing patterns composed of original wing cells and transdetermined wing cells in mosaics (GARCIA-BELLIDO, 1966b).

Recognition properties may also be at work during normal development *in situ*. Some of the late adjustments of patterns could be based on cell affinity properties of previously determined cells. The above described separation of clones in the dorsal and ventral surfaces whose cells contact along the wing margins, or of anterior and posterior clones along the vein 4 (p. 69), could be based on cell recognition properties. The clonal shift of the prospective sexcomb cells relative to the transverse row

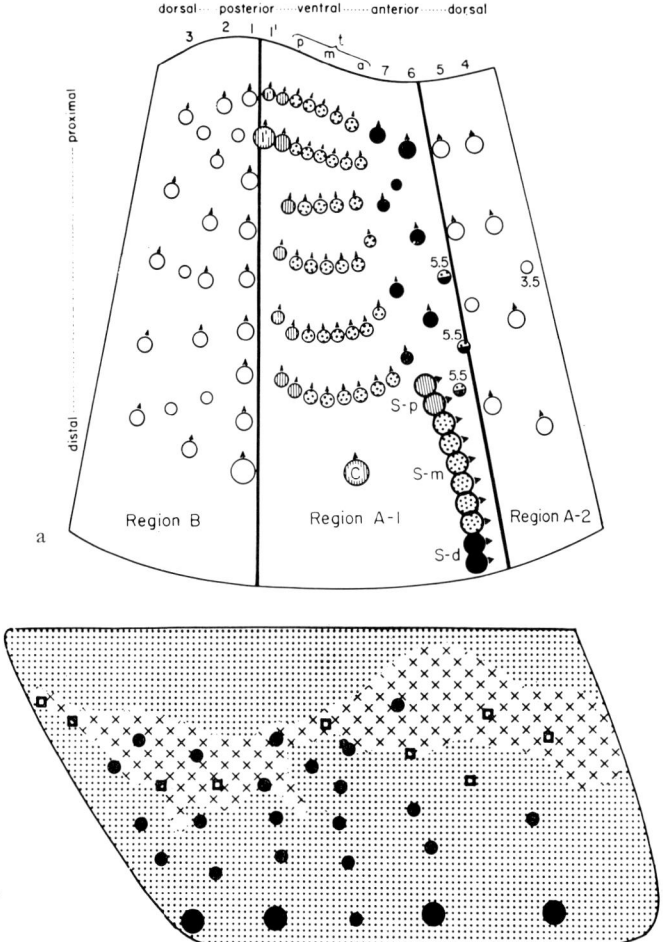

Fig. 7a, b. Morphogenetic movements preceding pattern distribution. a Cell lineage analysis of the male sexcomb. Observe the relative rotation of the three regions of the sexcomb ($s-p$, $s-m$, $s-d$) with respect to the transverse rows (t) of the same cell lineage (TOKUNAGA, 1962). b A clone of recombinant cells in the tergite marking the trichomes ($\times$) and the chaetes ($\bullet$ $\square$). Observe the location of a doubly mutant chaete ($\square$) outside the corresponding clone and of four normal chaetes ($\bullet$) within the mutant clone (GARCIA-BELLIDO and MERRIAM, 1971b)

cells to occupy the final position in the pattern (Fig. 7a) (TOKUNAGA, 1962) could also be based on specific cell recognition. The final location of the chaetes in a tergite pattern can also be explained by the selective migration of prospective chaete-forming cells. Clones of a genotype which simultaneously marks trichomes and chaetes show very frequently marked chaetes but not trichomes outside of the corresponding clone, or normal chaetes within marked clones (Fig. 7b) (GARCIA-BELLIDO and MERRIAM, 1971c).

C. Cell Interactions

The preceding analysis has revealed an extreme cell autonomy in the organization of the final pattern. Determination may precede the actual location of a cell in a preestablished position. We will now analyse some experiments which suggest the possibility that specific location, perhaps simultaneously with cell determination may be extrinsically imposed on the cells.

Specific *inductive mechanisms* acting from cell to cell can be invoked to explain the detailed organization of the final pattern. The bracts are epidermal cell modifications accompanying the chaetes in certain patterns of the legs (Fig. 1b—e). In reaggregates, they do not appear isolated, but always at the basis of their chaetes. They can be of a different genotype than that of the chaetes and so have a different cellular origin. Moreover, as seen before, bracts show always the same orientation as the corresponding chaetes. These facts taken together led us to assume that they are induced by the chaetes (GARCIA-BELLIDO, 1966a; TOBLER, 1966). This hypothesis is supported by further data. Treatment of imaginal disks with mitomycin C (TOBLER, 1969; TOBLER and PFLUGER, 1970), or with nitrogen mustard (TOBLER and MAIER, 1970) inhibits effectively the development of the tormogen cell of the chaete. All chaetes which lacked a socket were also deprived of the bract. In mutants where the chaetes are oriented in a disarranged fashion, as in *spiny legs (sple)* the bracts follow the orientation of the chaetes (GARCIA-BELLIDO, unpublished). Again, in another mutant, *shaven depilate (sv^{de})*, which removes the trichogen element but leaves the tormogen of many chaetes, bracts are absent in trichogen-less chaetes (NÖTHIGER, pers. comm.). Thus the presence of a normal developing chaete-organ, with polar singularities, is a prerequisit for the formation of a bract.

In combinates of leg and wing cells the bracts are always of the genotype of the leg cells, suggesting that wing cells are not competent to differentiate a bract under the induction of the chaete forming cells of the leg. This lack of induction is a general rule in heteronomous combinates, corroborated by the existence of faulty mosaics. Experiments, such as cell combinates, aimed at detecting assimilative or complementary inductions have either failed or have other possible interpretations (see GARCIA-BELLIDO, 1968a). However, for cells related topologically such as chaetes and bracts — and perhaps by cell lineage —, actual inductions could play a morphogenetic role. We do not know the actual mechanism of cell induction. Growing imaginal disks have intercellular membrane specialisations (see URSPRUNG, this volume). The density of such cell junctions is lower in young imaginal disks than in mature ones, and both much lower than in the adult epithelium (POODRY and SCHNEIDERMAN, 1970). We do not know the function of such cellular connections in imaginal disks. Some of them could provide directional channels for informational molecules to pass between neighbouring cells. If they were involved in pattern formation, one might expect them to be non-randomly distributed. This, however, has not yet been detected, nor is the developmental or functional stability of these membrane specialisations known.

The *organization of the cuticular elements in the tergites* in several insects is a recurrent subject of pattern analysis. Since the present status of the problem has been discussed in recent reviews (see LAWRENCE, 1970) we will restrict our analysis to *Drosophila*. Analyses of gynandromorphs and of mitotic recombination experiments show that the adult tergites derive from segmental bilateral groups of 8—10 imaginal cells, the histoblasts. These are set apart from the remaining larval epidermal cells, possibly at the blastoderm stage, and start dividing only after pupariation (GARCIA-BELLIDO and MERRIAM, 1971b). During the first 20 h after pupariation the proliferating histoblasts spread over the presumptive hemitergite area substituting the histolysing larval epidermal cells (ROBERTSON, 1936). Cauterization experiments in larvae indicate that the histoblasts of each hemitergite group in a single nest, equivalent to an imaginal disk, occupying a central position in the larval hemisegment (SANTAMARIA and GARCIA-BELLIDO, 1972). Pattern abnormalities have been found in mutant strains (SOBELS, 1952) and as a consequence of experimental injury (ZIMMERMANN, 1954; LÖBBECKE, 1958). Cauterization applied to different regions of each larval hemisegment produces characteristic abnormalities in the pattern of the adult tergite. These abnormalities correspond mainly to 1) reduction or increase in the normal amount of chaetes; 2) changes in the polarity and distribution of the adult elements, and 3) failures in the separation and correct connexion of neighbouring hemitergites. These damages are preferentially obtained by cauterization in the margins of the larval hemisegments, where no histoblasts are found. They cannot result from direct damage to the histoblast nest and, therefore, must derive from damage to the larval substratum upon which the proliferating histoblasts spread (SANTAMARIA and GARCIA-BELLIDO, 1972). Similar influences of the tergite margins and of neighbouring cells in the pattern and polarity of the tergite cells were experimentally observed in other insects, such as *Galleria* (PIEPHO, 1955; MARCUS, 1962; STUMPF, 1965), *Rhodnius* (LOCKE, 1959), *Oncopeltus* (LAWRENCE, 1966). However, in these insects the cuticular structures of the adult, or of the different larval instars, derive from the same epidermal cells, that is from cellular organizations laid down in the embryo. We do not know what is the nature of the positional signals originating in the substratum and reaching the invading cells in *Drosophila*. They could be of a diffusible nature, as suggested in other insects, or consist of cell to cell, or cell to substratum contact.

The *patterns of the wing veins* have been the subject of numerous experimental studies in *Drosophila* as well as in other insects (see HENKE, 1953). Embryological analyses of the formation of the wing veins in *Drosophila* indicate that they arise from lacunae separating both wing surfaces. These are detectable in mature larvae, and after a series of contraction and expansion processes form the definitive venation pattern in the 40 h old pupae (WADDINGTON, 1940). Several mutants are known which change the venation pattern. Some of them affect very precisely only certain regions of the venation system, and when subjected to genetic selection they can affect even more restricted portions of the pattern. That suggests rather complex mechanisms of vein morphogenesis (see WADDINGTON, 1962). The lack of regulation following experimental injuries in the larvae seems to indicate that the main venation pattern is laid down already in this developmental stage. However, in pupal stages minor alterations such as interrupted veins, or extra veins, are still possible (BRAUN, 1940; LEES, 1941). In these malformations the upper wing epithelium seems to lead, or induce, the lower wing (LEES, 1941). Interestingly, heat shocks in early pupal stages are able to produce venation changes which phenocopy known mutants (for review see HENKE, 1947; SCHATZ, 1951; STUMPF, 1959). Since in pupae the mitoses of the disk are coming to an end the obtained abnormal pattern cannot result from clonal changes but rather from experimental interference with the supracellular morphogenetic mechanisms.

In *Galleria* the lacunae system of the veins represents a channel through which tracheation and innervation occurs. Tracheation proceeds centrifugally from the wing hinge towards the margins, and innervation centripetally from the margins to the hinge. Experimental interference with the normal development shows that both tracheae and nerves follow the venation changes and thus are possibly inductively directed by the lacunae system (CLEVER, 1959).

Coordinate changes affecting simultaneously several cells seem to be responsable for the cuticular patterns arising after *regeneration and transdetermination*. KROEGER (1958b) sewed together anterior and posterior wing anlagen of mature *Ephestia* larvae by their hinge region and cultivated the organ combinates *in vivo*. After several days of culture both imaginal wings had formed a single chitinous epimeral complex composed of fractions of both anterior and posterior hinge patterns in different proportions, but organized in an integrated unity. Each fraction developed autonomously, differentiating the corresponding structures of its wing hinge. Apparently, regeneration — or cellular reorganization — occurred in the system as a whole, though their cells retained their previous anterior or posterior wing character. This example is reminiscent of the results of joining morphogenetic fields in amphibian embryos, or of the prepattern experiments mentioned on p. 71.

Regenerates of imaginal disk fragments normally show repetitions of the prospective structures present in the original fragment. Such structures are usually arranged into patterns of normal organization and size. In asymmetric disks, such as the leg disk, left halves produce left half duplications down to very delicate pattern details (NÖTHIGER and SCHUBIGER, 1966). It is possible that these duplications are the result of cell to cell heredity of the primitive determination followed by migration of the new cells to a new blastema, thus copying the previous cellular organization. However, it is known that regenerative growth takes place preferentially in the cells of the wound surface (KROEGER, 1958a, but see WILDERMUTH, 1968). In fact, a clonal analysis of the regenerates indicates that the newly formed part of the duplication does not result from strict cell heredity, nor from a reorganization of old and new cells (ULRICH, 1971; POSTLETHWAIT et al., 1971; NÖTHIGER and ULRICH, in prep.; see NÖTHIGER, this volume). Thus, cell arrangements can be imposed upon a population of proliferating cells by other cells already determined and organized in the old blastema. This is also the most reasonable interpretation of pattern duplications found in mosaics of autotypic and transdetermined (allotypic) cells (Fig. 8). Apparently, transdetermined cells may "copy" the pattern arrangements found in the territories to which they, due to the new cell affinities, attach.

The patterns of transdetermined territories could also arise by supracellular organising mechanisms. The normal sized and organized patterns of the transdetermined territories could derive from clonal regeneration, that is from a single original transdetermined cell. However, clonal analysis has shown that organized transdetermined territories in some cases derive from the descendents of more than one cell (GEHRING, 1967). We have seen that cells of normal mature imaginal disks, as a rule, do not change their determination in heteronomous combinates. However, it is not inconceivable that cells in regenerating blastemas are more ready to bend to assimilative induction. In this respect the finding that transdetermination occurs more readily in regenerating fragments of mature imaginal disks than of young disks is

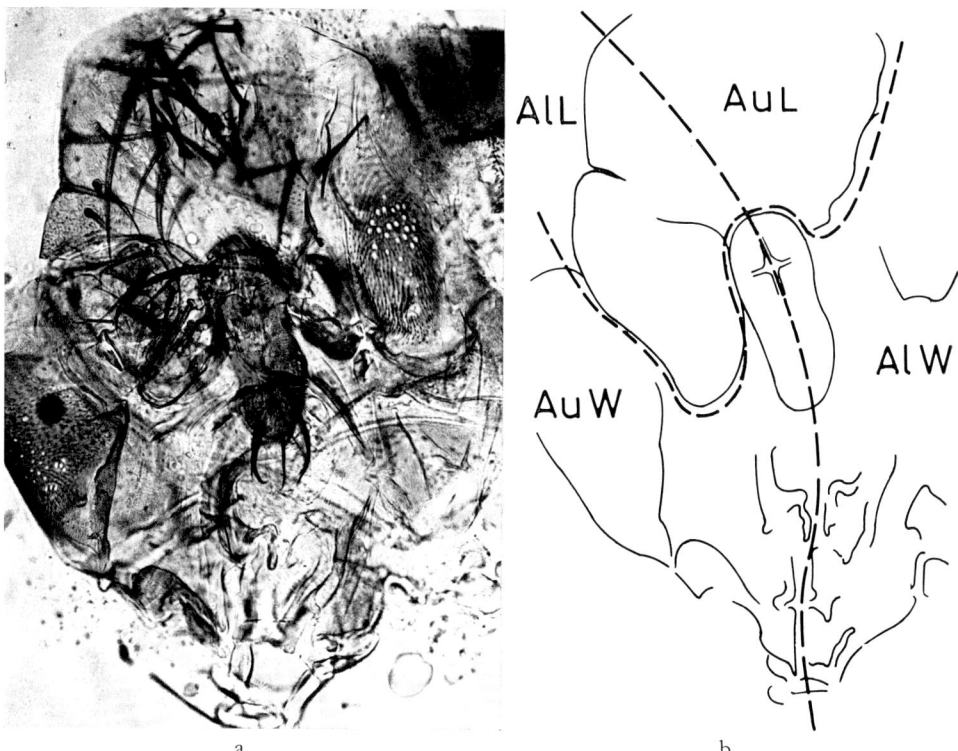

Fig. 8a, b. Pattern duplications in aggregates. Wing (*y* yellow; *mwh* multiple wing hairs). Cells were dissociated and mixed with leg (*e* ebony) cells and cultured 20 days in an adult fly. After metamorphosis autotypic *(Au)* and allotypic *(Al)* structures of leg *(L)* and wing *(W)* patterns appeared duplicated. a photograph, b schematic drawing

very illustrative (SCHUBIGER, 1968). As the author suggested the young imaginal disks have first to finish their normal program of larval development before their cells can take new developmental alternatives. Thus, any interpretation of cell behavior gained from regeneration experiments should only carefully be extended to cells during the normal development.

V. Concluding Remarks

An analysis of morphogenesis must explain how the genome acting within the cells controls multicellular organizations. To find the mechanisms leading to specific cell differentiation in specific positions of a pattern forms part of such an endeavour. The analysis of cuticular patterns in *Drosophila* has shown that specific cell arrangements are under genetic control down to minute details. Patterns seem to result from the superposition of various independent genetic processes, recognizable by the specific effects of many pattern mutations.

Patterns, considered as a whole, can be causally explained assuming that diffusible gene products reach, at different concentrations, different cells of the population. The different elements of a pattern would then correspond with as many singular points of conflict between several gradients acting simultaneously. Thus, individual cells would acquire their specific determination according to their position in the blastema ("positional information", WOLPERT, 1969). Pattern mutations could lead to the suppression — or addition — of diffusible gene products. Experiments have been performed to detect the influences of the cell population upon the single cells. Genetic mosaics as well as cell mixture experiments show, as a rule, an autonomous differentiation of the confronted cell types. In other examples, e.g. in the formation of tergite patterns and in regeneration experiments, though, localized cell differentiations seem to respond to supracellular organizations.

If the gene products are only active within the individual cells, pattern formation must be explained in other terms. It could be a direct consequence of the progressive segregation of cell lineages which would restrict both the relative location and the prospective differentiation of the proliferating cells. These restrictions could consist of the election of very specific developmental alternatives ("developmental pathways"; LEWIS, 1963), possibly corresponding to genetic decisions which would be prevented, shifted, or altered as a consequence of pattern mutations. The results of cell lineage studies in both normal and mutant patterns seem to fit in this explanation. Differential divisions or cell contact mediated inductions could represent the trigger mechanisms in these genetic alternatives.

We do not know the material bases of the signals arising from the blastema organization and reaching the single cells, nor do we know the nature and mechanisms of such clonal segregations. Certainly, further work in the genetic dissection of morphogenetic processes will add to an understanding of the language the single cells use to arrange into, and maintain, organized supracellular communities.

References*

AUERBACH, C.: The development of the legs, wings and halteres in wild type and some mutant strains of *Drosophila melanogaster*. Trans. Roy. Soc. Edinburgh. **58**, 787–815 (1936).

BECKER, H. J.: Über Röntgenmosaikflecken und Defektmutationen am Auge von *Drosophila* und die Entwicklungsphysiologie des Auges. Z. Vererbl. **88**, 333–373 (1957).

*BODENSTEIN, C.: The postembryonic development of *Drosophila*. In: DEMEREC, M. (ed.): Biology of Drosophila, pp. 275–357. New York: Wiley 1950.

BODENSTEIN, C., ABDEL-MALEK, A.: The induction of *aristapedia* by nitrogen mustard in *Drosophila virilis*. J. Exptl. Zool. **111**, 95–115 (1949).

BRAUN, W.: The effect of punctures in the developing wings of several mutants of *Drosophila melanogaster*. J. Exptl. Zool. **84**, 325–349 (1940).

BRYANT, P. J.: Cell lineage relationships in the imaginal wing disc of *Drosophila melanogaster*. Develop. Biol. **22**, 389–411 (1970).

BRYANT, P. J., SCHNEIDERMAN, H. A.: Cell lineage, growth and determination in the imaginal leg disc of *Drosophila melanogaster*. Develop. Biol. **20**, 263–290 (1969)

---

* Review articles on pattern formation are marked with an asterisk.

CHAN, L., GEHRING, W.: Determination of blastoderm cells in *Drosophila melanogaster*. Proc. Nat. Ac. Sci. USA. **68**, 2217—2221 (1971).

CLEVER, U.: Über experimentelle Modifikationen des Geäders und die Beziehung zwischen dem Versorgungssystem im Schmetterlingsflügel. Untersuchungen an *Galleria mellonella*. Wilhelm Roux' Archiv. **151**, 242—279 (1959).

CURTIS, A. S. G.: The cell surface: its molecular role in morphogenesis. Logos Press. London (1967).

DATTA, R. K., MUKHERJEE, A. S.: Developmental genetics of the mutant *combgap* in *Drosophila melanogaster*. I. Effect on the morphology and chaetotaxy of the prothoracic leg. Genetics **68**, 269—286 (1971).

DUBININ, N. P.: Allelomorphentreppen bei *Drosophila melanogaster*. Biol. Zentblt. **49**, 328—339 (1929).

FRISTROM, D.: Cellular degeneration in the production of the some mutant phenotypes in *Drosophila melanogaster*. Mol. Gen. Genet. **103**, 363—379 (1969).

GARCIA-BELLIDO, A.: Larvalentwicklung transplantierter Organe von *Drosophila melanogaster* im Adultmilieu. J. Ins. Physiol. **11**, 1071—1078 (1965).

GARCIA-BELLIDO, A.: Pattern reconstruction by dissociated imaginal disc cells of *Drosophila melanogaster*. Develop. Biol. **14**, 278—306 (1966a).

GARCIA-BELLIDO, A.: Changes in selective affinity following transdetermination in imaginal disk cells of *Drosophila melanogaster*. Exptl. Cell. Res. **44**, 382—397 (1966b).

GARCIA-BELLIDO, A.: Histotypic reaggregation of dissociated imaginal disc cells of *Drosophila melanogaster* cultured in vivo. Wilhelm Roux' Arch. **158**, 211—217 (1967).

GARCIA-BELLIDO, A.: Cell affinities in antennal homoeotic mutants of *Drosophila melanogaster*. Genetics **59**, 487—499 (1968a).

GARCIA-BELLIDO, A.: Cell lineage in the wing disc of *Drosophila melanogaster*. Genetics **60**, 181 (1968b) (Abstr.).

GARCIA-BELLIDO, A., MERRIAM, J. R.: Bristles or hairs: Heredity of a genetic decision in *Drosophila melanogaster*. Proc. Nat. Acad. Sci. USA (Abstr.) **61**, 1147 (1968).

GARCIA-BELLIDO, A., MERRIAM, J. R.: Cell lineage of the imaginal discs in *Drosophila* gynandromorphs. J. Exptl. Zool. **170**, 61—76 (1969).

GARCIA-BELLIDO, A., MERRIAM, J. R.: Parameters of the wing imaginal disc development of *Drosophila melanogaster*. Develop. Biol. **24**, 61—87 (1971a).

GARCIA-BELLIDO, A., MERRIAM, J. R.: Genetic analysis of cell heredity in imaginal discs of *Drosophila melanogaster*. Proc. Nat. Acad. Sci. USA **68**, 2222—2226 (1971b).

GARCIA-BELLIDO, A., MERRIAM, J. R.: Clonal parameters of tergite development in *Drosophila*. Develop. Biol. **26**, 264—276 (1971c).

GARCIA-BELLIDO, A., SANTAMARIA, P.: Developmental analysis of the wing disc in the mutant *engrailed* of *Drosophila melanogaster*. Genetics (in press).

GEHRING, W.: Phenocopies produced by 5-fluoro-uracil. Dros. Inf. Serv. **39**, 102 (1964).

GEHRING, W.: Übertragung und Änderung der Determinationsqualitäten in Antennenscheiben-Kulturen von *Drosophila melanogaster*. J. Embryol. exp. Morph. **15**, 77—111 (1966).

GEHRING, W.: Clonal analysis of determination dynamics in cultures of imaginal disks in *Drosophila melanogaster*. Develop. Biol. **18**, 438—456 (1967).

GEHRING, W.: The stability of the determined state in cultures of imaginal disks of *Drosophila*; this volume.

*GEHRING, W., NÖTHIGER, R.: The imaginal discs of *Drosophila*. In: WADDINGTON, C. H., COUNCE, S. J., (Eds.): Developmental systems: Insects. New York: Academic Press (in press).

GLOOR, H.: Phänokopie-Versuche mit Äther in Drosophila. Rev. suisse Zool. **54**, 637—712 (1947).

GOTTLIEB, F. J.: Genetic control of pattern determination in *Drosophila*. The action of Hairy-wing. Genetics **49**, 739—760 (1964).

*HADORN, E.: Regulation and differentiation within field-districts in imaginal discs of *Drosophila*. J. Embryol. exp. Morph. **1**, 213—216 (1953).

HADORN, E.: Konstanz, Wechsel und Typus der Determination und Differenzierung in Zellen aus männlichen Genitalanlagen von *Drosophila melanogaster* nach Dauerkultur *in vivo*. Develop. Biol. **13**, 424—509 (1966a).

HADORN, E.: Über eine Änderung musterbestimmender Qualitäten in einer Blastemkultur von *Drosophila melanogaster*. Rev. suisse Zool. **73**, 253—265 (1966b).

HADORN, E., ANDERS, G., URSPRUNG, H.: Kombinate aus teilweise dissoziierten Imaginalscheiben verschiedener Mutanten und Arten von *Drosophila*. J. exp. Zool. **142**, 159—175 (1959).

HADORN, E., BERTANI, G., GALLERA, J.: Regulationsfähigkeit und Feldorganisation der männlichen Genital-Imaginalscheibe von *Drosophila melanogaster*. Wilhelm Roux' Arch. Entwickl.-Mech. Org. **144**, 31—70 (1949).

HADORN, E., BUCK, D.: Über Entwicklungsleistungen transplantierter Teilstücke von Flügel-Imaginalscheiben von *Drosophila melanogaster*. Rev. suisse Zool. **69**, 302—310. (1962).

HADORN, E., GLOOR, H.: Transplantationen zur Bestimmung des Anlagemusters in der weiblichen Genital-Imaginalscheibe von *Drosophila*. Rev. suisse Zool. **53**, 495—501 (1946).

HANNAH-ALAVA, A.: Morphology and Chaetotaxy of the Legs of *Drosophila melanogaster*. J. Morph. **103**, 281—310 (1958).

*HENKE, K.: Einfache Grundvorgänge in der tierischen Entwicklung. II. Naturwissenschaften **34**, 149—157, 180—186 (1947).

*HENKE, K.: Die Hauptformen der Gliederungsvorgänge in der Entwicklung des Insektenflügels. Verh. dtsch. Zool. Ges. Wilhelmshaven 42—52 (1951).

*HENKE, K.: Die Musterbildung der Versorgungssysteme im Insektenflügel. Biol. Zbl. **72**, 1—51 (1953).

KROEGER, H.: Über Doppelbildungen in die Leibeshöhle verpflanzter Flügelimaginalscheiben von *Ephestia kühniella*. Wilhelm Roux' Arch. Entwickl.-Mech. Org. **150**, 401—424 (1958a).

KROEGER, H.: Determinationsmosaike aus kombiniert implantierten Imaginalscheiben von *Ephestia kühniella*. Zeller. Wilhelm Roux' Arch. Entwickl.-Mech. Org. **151**, 113—135 (1958b).

KROEGER, H.: The genetic control of genital morphology in *Drosophila*. Wilhelm Roux Arch. Entwickl.-Mech. Org. **151**, 301—322 (1959).

*KROEGER, H.: Die Entstehung von Form im morphogenetischen Feld. Naturwissenschaften **47**, 148—153 (1960).

KURODA, Y.: Characteristic aggregation pattern of dissociated imaginal disc cells of *Drosophila melanogaster* larvae in rotation culture. Dros. Inf. Serv. **44**, 109 (1969).

LAWRENCE, P. A.: Gradients in the insect segment: The orientation of hairs in the milkweed bug, *Oncopeltus fasciatus*. J. exp. Biol. **44**, 607—620 (1966).

*LAWRENCE, P. A.: Polarity and patterns in the postembryonic development of insects Advanc. Ins. Physiol. **7**, 197—266 (1970).

LEES, A. D.: Operations in the pupal wing of *Drosophila melanogaster*. J. Genet. **42**, 115—142 (1941).

LEES, A. D., WADDINGTON, C. H.: The development of the bristles in normal and some mutant types of *Drosophila melanogaster*. Proc. roy. Soc. London Ser. B **131**, 87—110 (1942).

LESSEPS, R. J.: Culture of dissociated *Drosophila* embryos: Aggregated cells differentiate and sort-out. Science **148**, 502—503 (1965).

LEVINE, M.: Phage morphogenesis. Ann. Rev. Genet. **3**, 323—342 (1969).

*LEWIS, E. B.: Genetic control and regulation of developmental pathways. In: LOCKE, M. (Ed.): Role of chromosomes in development, pp. 232—251. New York: Academic Press 1964.

*LEWIS, E. B.: Genes and developmental pathways. Amer. Zoologist. **3**, 33—56 (1963).

LEWIS, E. B.: Proc. XIIth Int. Cong. Genet. Tokyo (1968).

LINDSLEY, D. L., GRELL, E. H.: Genetic variations in *Drosophila melanogaster*. Carnegie Inst. Wash. Publ. No. **627** (1968).

LÖBBECKE, E. A.: Über die Entwicklung der imaginalen Epidermis des Abdomens von *Drosophila*, ihre Segmentierung und die Determination der Tergite. Biol. Zbl. **77**, 209—237 (1958).

Locke, M.: The cuticular pattern in an insect, *Rhodnius prolixus*. Stal. J. exp. Biol. **36**, 459–478 (1959).

Lüönd, H.: Untersuchungen zur Mustergliederung in fragmentierten Primordien des männlichen Geschlechtsapparates von *Drosophila séguyi*. Develop. Biol. **3**, 615–656 (1961).

Marcus, W.: Untersuchungen über die Polarität der Rumpfhaut von Schmetterlingen. Wilhelm Roux' Arch. Entwickl.-Mech. Org. **154**, 56–102 (1962).

Maynard Smith, J., Sondhi, K. C.: The genetics of a pattern. Genetics **45**, 1039–1050 (1960).

Murphy, C.: Determination of the dorsal mesothoracic disc in *Drosophila*. Develop. Biol. **15**, 368–394 (1967).

Murphy, C., Tokunaga, C.: Cell lineage in the dorsal mesothoracic disc of *Drosophila*. J. exp. Zool. **175**, 197–220 (1970).

Nöthiger, R.: Differenzierungsleistungen in Kombinaten, hergestellt aus Imaginalscheiben verschiedener Arten, Geschlechter und Körpersegmente von *Drosophila*. Wilhelm Roux' Arch. Entwickl.-Mech. Org. **155**, 269–301 (1964).

*Nöthiger, R.: The larval development of imaginal disks; this volume.

Nöthiger, R., Schubiger, G.: Developmental behaviour of fragments of symmetrical and asymmetrical imaginal discs of *Drosophila melanogaster*. J. Embryol. exp. Morphol. **16**, 355–368 (1966).

Nöthiger, R., Ulrich, E.: Cell lineage and determination in the male genital disk of *Drosophila melanogaster*. (in prep.).

Piepho, H.: Über die Ausrichtung der Schuppenbalge und Schuppen am Schmetterlingsrumpf. Naturwissenschaften **42**, 22 (1955).

Poodry, C. A., Bryant, P. J., Schneiderman, H. A.: The mechanism of pattern reconstruction by dissociated imaginal discs of *Drosophila melanogaster*. Develop. Biol. **26**, 464–477 (1971).

Poodry, C. A., Schneiderman, H. A.: The ultrastructure of the developing leg of *Drosophila melanogaster*. Wilhelm Roux' Arch. Entwickl.-Mech. Org. **166**, 1–44 (1970).

Postlethwait, J. H., Poodry, C. A., Schneiderman, H. A.: Cellular dynamics of pattern duplication in imaginal discs of *Drosophila melanogaster*. Develop. Biol. **26**, 125–132 (1971).

Postlethwait, J. H., Schneiderman, H. A.: A clonal analysis of determination in *Antennapedia*, a homoeotic mutant of *Drosophila melanogaster*. Proc. nat. Acad. Sci. (Wash.) **64**, 176–183 (1969).

Postlethwait, J. H., Schneiderman, H. A.: A clonal analysis of development in *Drosophila melanogaster*: Morphogenesis, determination and growth in the wild type antenna. Develop. Biol. **24**, 477–519 (1971).

Ripoll, P.: The embryonic organization of the imaginal wing disc of *Drosophila melanogaster*. Wilhelm Roux' Arch. Entwickl.-Mech. Org. **169**, 200–215 (1972).

Roberts, P.: Mosaics involving *aristapedia*, a homoeotic mutant of *Drosophila melanogaster*. Genetics **49**, 593–598 (1964).

Robertson, C. W.: The metamorphosis of *Drosophila melanogaster*, including an accurately timed account of the principal morphological changes. J. Morphol. **59**, 351–400 (1936).

Santamaria, P., Garcia-Bellido, A.: Localization and growth pattern of the tergite anlage of *Drosophila*. Submitted to J. Embryol. exp. Morph.

Schatz, E.: Über die Formbildung der Flügel bei Hitzemodifikationen und Mutationen von *Drosophila melanogaster*. Biol. Zbl. **70**, 305–353 (1951).

*Schneiderman, H. A., Bryant, P. J.: Genetic analysis of developmental mechanisms in *Drosophila*. Nature (Lond.) **234**, 187–194 (1971).

Schubiger, G.: Anlageplan, Determinationszustand und Transdeterminationsleistungen der männlichen Vorderbeinscheibe von *Drosophila melanogaster*. Wilhelm Roux' Arch. Entwickl.-Mech. Org. **160**, 9–40 (1968).

Schubiger, G., Schubiger, M., Hadorn, E.: Mischungsversuche mit Keimteilen von *Drosophila melanogaster* zur Ermittlung des Determinationszustandes imaginaler Blasteme im Embryo. Wilhelm Roux' Arch. Entwickl.-Mech. Org. **163**, 33–39 (1969).

Sobels, F. H.: Genetics and Morphology of the Genotype "Asymmetric" with special reference to its "Abnormal abdomen" character (*Drosophila melanogaster*). Genetics **26**, 117—279 (1952).

Sondhi, K. C.: The evolution of a pattern. Evolution **16**, 186—191 (1962).

*Sondhi, K. C.: The biological foundations of animal patterns. Quart. Rev. Biol. **38**, 289—327 (1963).

Spreij, T. E.: Cell death during the development of the imaginal disks of *Calliphora erythrocephala*. Neth. J. Zool. **21**, 221—264 (1971).

Steinberg, M. S.: Does differential adhesion govern selfassembly processes in histogenesis? Equilibrium configurations and the emergence of a hierarchy among populations of embryonic cells. J. exp. Zool. **173**, 395—434 (1970).

Stern, C.: The prospective significance of imaginal discs of *Drosophila*. J. Morphol. **67**, 107—122 (1940).

*Stern, C.: Genes and developmental patterns. Caryologia Suppl. 6, **1**, 355—369 (1954).

Stern, C.: Entwicklung und die Genetik von Mustern. Naturwissenschaften **52**, 357—365 (1965).

Stern, C., Hannah, A.: The sex-combs in gynanders of *Drosophila melanogaster*. Portugal. Acta Biol. Ser. A (R. B. Goldschmidt vol.) 798—812 (1950).

Stern, C., Tokunaga, C.: Non-autonomy in differentiation of pattern-determining genes in *Drosophila*. I. The sexcomb of *eyeless-dominant*. Proc. nat. Acad. Sci. (Wash.) **57**, 658—664. (1967).

Stumpf, H.: Die Richtungen der Teilungsspindeln auf dem Puppenflügel von *Drosophila* im Verlaufe der Mitosenperiode. Biol. Zbl. **75**, 17—27 (1956).

Stumpf, H.: Die Wirkung von Hitzereizen auf Entwicklungsvorgänge im Puppenflügel von *Drosophila*. Biol. Zbl. **78**, 116—142 (1959).

Stumpf, H.: Deutung der Richtungsmuster der Schuppen von *Galleria mellonella* auf Grund eines Konzentrationsgefälles. Naturwissenschaften **52**, 522 (1965).

Sturtevant, A. H.: The claret mutant type of *Drosophila simulans*, a study of chromosome elimination and cell-lineage. Z. wiss. Zool. **135**, 323—356 (1929).

Sturtevant, A. H.: Studies on the bristle pattern of *Drosophila*. Develop. Biol. **21**, 48—61 (1970).

Tobler, H.: Zellspezifische Determination und Beziehung zwischen Proliferation und Transdetermination in Bein- und Flügelprimordien von *Drosophila melanogaster*. J. Embryol. exp. Morph. **16**, 609—633 (1966).

Tobler, H.: Beeinflussung der Borstendifferenzierung und Musterbildung durch Mitomycin bei *Drosophila melanogaster*. Experientia (Basel) **25**, 213—214 (1969).

Tobler, H., Pfluger, M.: Untersuchung zur Wirkung von Mitomycin C auf die Entwicklung der männlichen Vorderbeinscheibe und die Differenzierung des Borstenorgans von *Drosophila melanogaster* nach Transplantation in larvale Wirte. Wilhelm Roux' Arch. Entwickl.-Mech. Org. **164**, 293—302 (1970).

Tobler, H., Maier, V.: Zur Wirkung von Senfgaslösungen auf die Differenzierung des Borstenorganes und auf die Transdeterminationsfrequenz bei *Drosophila melanogaster*. Wilhelm Roux' Arch. Entwickl.-Mech. Org. **164**, 303—312 (1970).

Tokunaga, C.: Cell lineage and differentiation in the male foreleg of *Drosophila melanogaster*. Develop. Biol. **4**, 485—516 (1962).

Townes, P. L., Holtfreter, J.: Directed movements and selective adhesion of embryonic amphibian cells. J. exp. Zool. **128**, 53—120 (1955).

Ulrich, E.: Cell lineage, Determination und Regulation in der weiblichen Genitalimaginalscheibe von *Drosophila melanogaster*. Wilhelm Roux' Arch. Entwickl.-Mech. Org. **167**, 64—82 (1971).

Ursprung, H.: Fragmentierungs- und Bestrahlungsversuche zur Bestimmung von Determinationszustand und Anlageplan der Genitalscheiben von *Drosophila melanogaster*. Wilhelm Roux' Arch. Entwickl.-Mech. Org. **151**, 504—558 (1959).

*Ursprung, H.: Development and Genetics of patterns. Amer. Zoologist. **3**, 71—86 (1963).

Ursprung, H.: The formation of patterns in development. In: Locke, M. (Ed.): Major Problems in Developmental Biology, pp. 177—216. New York: Academic Press 1967.

*Ursprung, H.: The fine structure of imaginal disks; this volume.

URSPRUNG, H., HADORN, E.: Weitere Untersuchungen über Musterbildung in Kombinaten aus teilweise dissoziierten Flügelimaginalscheiben von *Drosophila melanogaster*. Develop. Biol. **4**, 40–66 (1962).

WADDINGTON, C. H.: The genetic control of wing development in *Drosophila*. J. Genet. **41**, 75–139 (1940).

WADDINGTON, C. H.: Some developmental effects of x-rays in *Drosophila*. J. exp. Biol. **19**, 101–117 (1942).

*WADDINGTON, C. H.: New patterns in genetics and development. New York: Columbia Univ. Press 1962.

WILDERMUTH, H.: Autoradiographische Untersuchungen zum Vermehrungsmuster der Zellen in proliferierenden Rüsselprimordien von *Drosophila melanogaster*. Develop. Biol. **18**, 1–13 (1968).

*WOLPERT, L.: Positional information and the spatial pattern of cellular differentiation. J. theor. Biol. **25**, 1–47 (1969).

ZALOKAR, M.: L'ablation des disques imaginaux chez la larve de *Drosophila*. Rev. suisse Zool. **50**, 232–237 (1943).

ZIMMERMANN, W.: Über genetisch und modifikatorisch bedingte Störungen der Segmentierung bei *Drosophila melanogaster*. Z. Vererbungsl. **86**, 327–372 (1954).

The Fine Structure of Imaginal Disks

HEINRICH URSPRUNG

Zoologisches Institut, Labor für Entwicklungsbiologie, ETH, Zürich

I. Introduction

Answers to the following questions were expected from fine structure analyses of imaginal disks:

1) How are disks organized at the cellular level? Specifically, how many cell layers and cell types constitute a disk?

2) Is the determined state, for whose exploration imaginal disks appear so suited, expressed morphologically? Are the cells of a wing disk, e.g., distinguishable from those of an eye disk at the fine structural level? Or are cells of known lineage within a disk distinguishable from their neighbors? And finally, is transdetermination expressed at the morphological level?

3) How do the cells of imaginal disks, derivatives of the larval hypodermis, respond to the humoral control that is known to govern the larval molts?

4) How do the internally located disks form the exterior adult appendages during morphogenesis?

5) Is there a morphological parallel to the cell-cell communication that might convey positional information when the adult pattern of surface specializations is laid down?

6) Does selective cell death occur during disk morphogenesis?

Table 1

Disk	Organism	Reference
Antenna, wing	Calliphora erythrocephala	AGRELL (1966)
Wing	Calliphora erythrocephala	AGRELL (1968)
Leg	Sarcophaga bullata	CHIARODO and DENYS (1968)
Wing	Drosophila melanogaster	FRISTROM (1968)
Wing, eye	Drosophila melanogaster	FRISTROM (1969)
Male genital	Drosophila melanogaster	URSPRUNG and SCHABTACH (1968)
Wing	Drosophila melanogaster	URSPRUNG and SCHABTACH (1972)
Wing	Drosophila melanogaster	WEHMAN (1969), WEHMAN and BRAGER (1971)
Leg	Drosophila melanogaster	POODRY and SCHNEIDERMAN (1970)
Eye	Drosophila melanogaster	PERRY (1968)
Eye	Drosophila melanogaster	WADDINGTON and PERRY (1960)
Eye	Drosophila melanogaster	GATEFF et al. (1969)

7) Is the development of disks truly autonomous, or are cells or materials taken up from the environment?

8) Are mutations that are known to affect disk morphogenesis, visible at the morphological level?

Most papers published on imaginal disk fine structure have addressed themselves to one or more of these questions. Table 1 gives a survey of these analyses. For brevity, individual references will not be given in the remainder of this review except when specific comparisons are made or when isolated observations are reported.

II. Fine Structural Organization of Late Third Instar Larval Imaginal Disks

At first sight a typical disk appears as a vesicle built of an epithelium thrown into folds on one side (the disk proper) and covered on the other side by a smooth peripodial membrane continuous with the folds; the entire vesicle is surrounded by a noncellular basement lamina (Fig. 1).

Closer inspection shows four different cell types to occur in disks: true epithelial cells that make up the disk proper and the peripodial membrane, adepithelial cells, tracheolar cells, and nerve cells. The three latter cell types are found in the space between the epithelial cells of the disk proper and the basement lamina (Figs. 1 and 2).

There is compelling evidence that the epithelial cells during metamorphosis give rise to the ectodermal structures derived from the disks. By contrast, the role of the adepithelial cells has been a matter of controversy. Variably called "mesenchyme", "mesodermal", or "mesoblast" cells by light microscopists, these cells are believed to give rise to adult musculature. But the evidence for this is indirect yet.

No disk has been reconstructed from serial sections in its entirety. Because of the folded nature of the epithelium, it is difficult to say with certainty that it is single-layered, although this is seen to be the case for extensive regions of the disks including the peripodial membrane. Most likely, the seemingly multilayered epithelial portions represent a singlelayered zone sectioned tangentially. I think it is safe to say, therefore, that the true epithelial portions of disks are essentially singlelayered.

The epithelial cells of the late third instar disk proper are columnar (Figs. 1 and 2). Some 20—30 μ high, they have a strong polarity, with a border of microvilli at the apical side facing the lumen of the disk. Their bases are facing the layer(s) of adepithelial cells where these are present. Except for shape differences depending on the location in grooves or folds, the fine structure of all columnar cells is identical (Fig. 3). The cytoplasm is packed full with free ribosomes. Very little endoplasmic reticulum (ER) is present, and one sees but an occasional profile of rough ER. Golgi bodies, coated vesicles, multivesicular bodies, lipid droplets, mitochondria are present. Quite frequently, microtubules are seen arranged in the longitudinal axis of these cells. Microtubules have been seen to project into microvilli, an observation consistent with the assumed cytoskeletal function of these specializations (WEHMAN, 1969).

This same fine structure is again encountered in the epithelial cells of the peripodial membrane. These cells are shorter than the columnar cells of the disk proper, and their basal surfaces rest directly upon the basement lamina.

In contrast to these true epithelial cells, the adepithelial cells of wing and leg disks and the cells of the exterior epithelium in genital disks have no microvilli, nor pronounced polarity. They are not columnar, but cuboidal or squamous. Their long axis if detectable is perpendicular to that of the true epithelial cells. In genital disks, these cells appear organized into a separate tissue that is patched onto the epithelial layer (Fig. 1). In wing and leg disks, they are arranged in much looser fashion (Fig. 2). Some were seen to have filopodial extensions which might indicate that they migrate.

Fig. 1. Parasagittal section, about $1/4$ from the lateral tip, of a male genital disk dissected from a larva 104 h after oviposition. The picture is a montage of 29 electron micrographs of a single section. Lead citrate stained. Mag: × 1800. *CE* columnar epithelium; *EE* exterior epithelium; *IE* interior epithelium; *L* lumen; *MV* microvilli; *NCE* non-cellular envelope; *T* trachea. The black streaks are folds in the section. From URSPRUNG and SCHABTACH (1968)

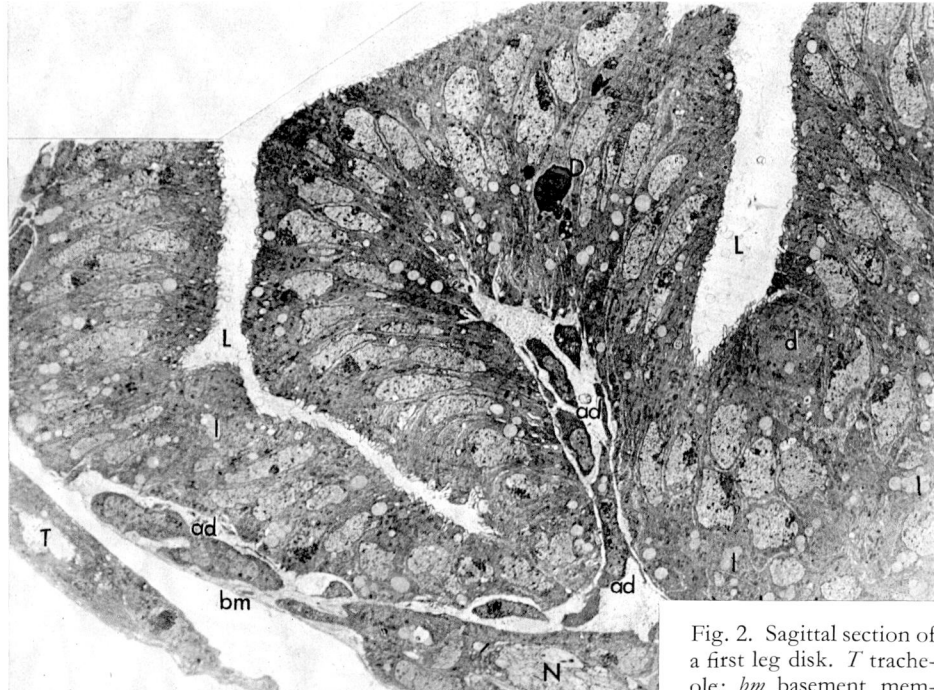

Fig. 2. Sagittal section of a first leg disk. *T* tracheole; *bm* basement membrane; *N* nerve; *ad* adepithelial cells; *L* lumen; *D* dead cell; *l* lipid droplets; *d* dividing cell. ×1465 (From POODRY and SCHNEIDERMAN, 1970)

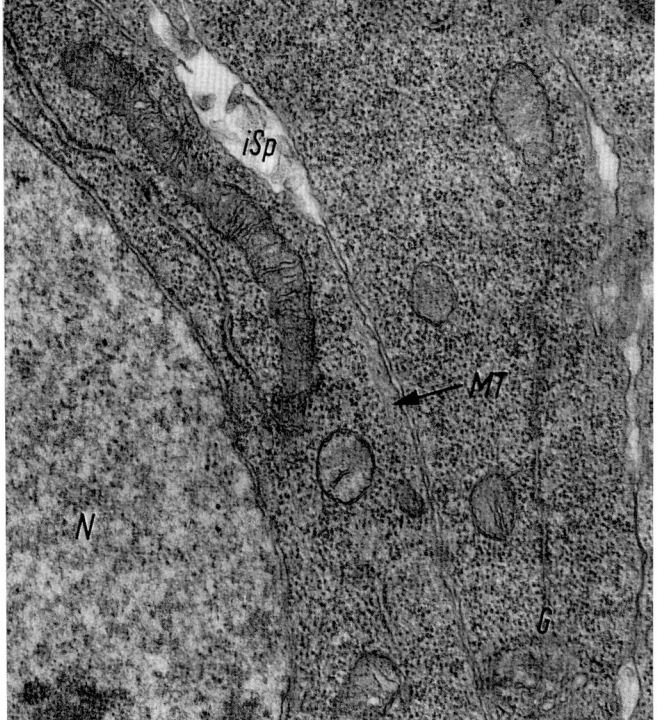

Fig. 3. Male genital disk epithelial cell fixed 75 min after puparium formation. Note abundance of free ribosomes. *G* Golgi; *iSp* intercellular space; *MT* microtubules; *N* nucleus. ×33600 (From URSPRUNG and SCHABTACH, 1968)

Gap junctions have been seen to connect individual adepithelial cells to cells of the columnar epithelium. Within the mass of adepithelial cells, nerve cells have been observed, and within the exterior epithelium, tracheoles (Figs. 1 and 2).

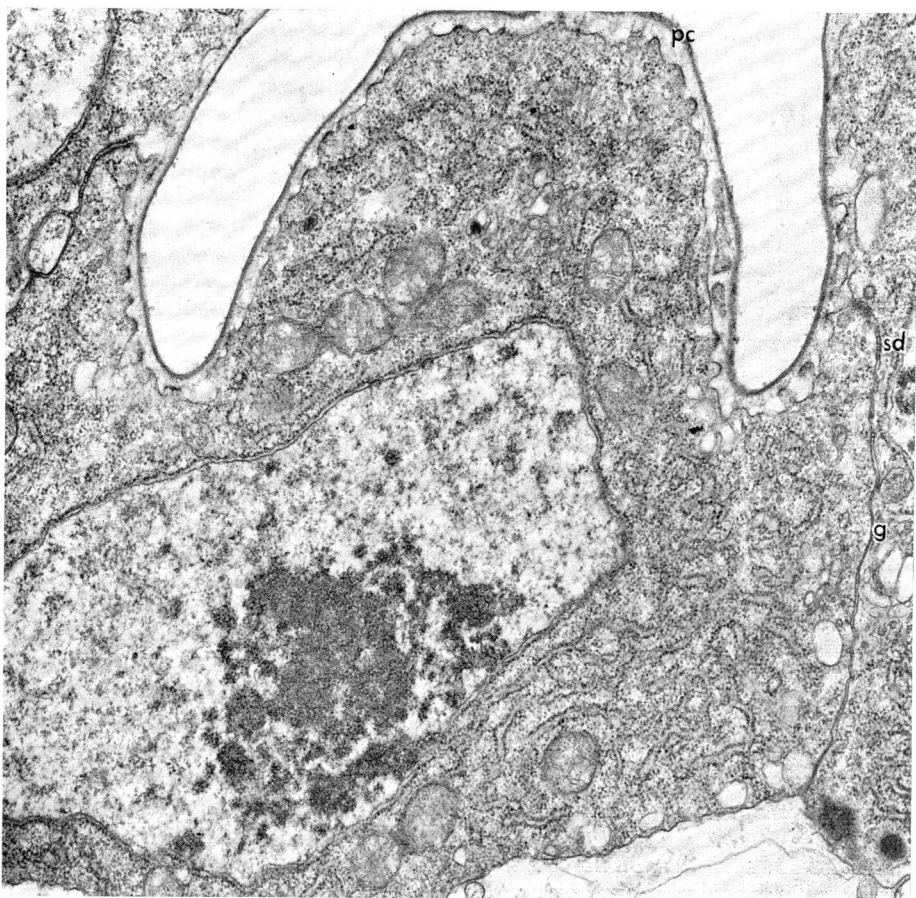

Fig. 4. Pupal leg cell. Note profiles of rough ER, and cuboidal shape of cell. *g* gap junction; *pc* pupal cuticle; *sd* septate desmosome. × 18000 (From POODRY and SCHNEIDERMAN, 1970)

Isolation and microinjury experiments (see NÖTHIGER, this volume) clearly indicate a rigid determination of cell groups or even single cells within imaginal disks. But the morphological basis of this condition has not been revealed by the electron microscope studies reported thus far. It is easy to recognize a particular disk by its arrangement of folds and grooves; but the different disks cannot be told apart on the basis of fine structure of individual cells.

Two possible exceptions should be mentioned. One, the columnar cells of the wing-forming area of *Drosophila* wing disks according to FRISTROM (1969) can be told apart from the cells of other regions of the disk on the basis of their lipid content. Since lipid droplets have been seen in other disks too, and since their frequency varies along the time axis of development (see III), I doubt that they are reliable criteria for cell identification. Two, the retinular cells of the white prepupa that has just formed the puparium are already grouped into bundles of eight, and are individually characterized by their spindle-shape (PERRY, 1968). We do not know whether they would maintain this shape upon isolation, however.

On the basis of the available data, I doubt very much that origin or fate of disk derived cell lines as they are used in transdetermination experiments (see GEHRING, this volume) can be diagnosed by electron microscopy.

III. Developmental Changes in Disk Fine Structure

A. Epithelial Cells

In late second and early third instar disks, and again after puparium formation, the epithelial cells of the disk proper are not columnar, but rather cuboidal. They carry fewer microvilli than later. They contain little if any ER, and it is of the smooth type. Towards the end of the last larval instar, the incidence of rough ER increases. Within an hour or two after puparium formation, the fine structure of the columnar cells in genital and leg disks changes dramatically in that extensive areas of rough ER appear, but free ribosomes nevertheless remain abundant (Fig. 4). The incidence of Golgi bodies increases, and lipid droplets become more numerous and later are seen to coalesce into large drops within which glycogen has been seen. Multivesicular bodies have been seen to fuse with coated vesicles. Fewer microtubules are seen in pupal cells than earlier. In wing disks of *Calliphora*, an increase in the frequency of nuclear protrusions has been reported to occur towards the end of the last larval instar (AGRELL, 1968). In *Sarcophaga* leg disks, enlargement of the nucleolus was seen towards the end of the third instar (CHIARODO and DENYS, 1968). A very striking developmental change is the deposition of cuticle, apparent on the microvillar border for the first time at puparium formation. Prior to this stage, no evidence of microvillar cuticle synthesis has been found. Specifically, wing disks apparently do not respond with cuticle synthesis to the hormonal signals that cause the molt from the second to the third larval instar (WEHMAN, 1969). Although material is occasionally seen in the lumina of all disks looked at, its origin is uncertain, and it does not seem to be a remnant of larval molts of disks. At puparium formation (120 h after oviposition), a "fuzz" is seen to connect the tips of the microvilli. Some five hours later, cuticulin appears, and after one more hour, epicuticle is plainly visible. By eight hours after puparium formation, endocuticle begins to be laid down, which at 132 h after oviposition consists of 5—7 layers (Fig. 5).

This is not the place to go into the controversy of nomenclature of these layers. What should be discussed however is whether the cuticle laid down between puparium formation and pupation, which occurs at 132 h after oviposition, should be called "prepupal" (WEHMAN, 1969) or "pupal" (POODRY and SCHNEIDERMAN, 1970). The latter authors have produced an electron micrograph of leg disks showing that after completion of the cuticle a layer appears between it and the inner surface of the puparium. They logically conclude that this extra layer, appearing last, cannot be a prepupal cuticle, but is the innermost layer of the puparium, split off by some process. Their electron micrograph supports this view.

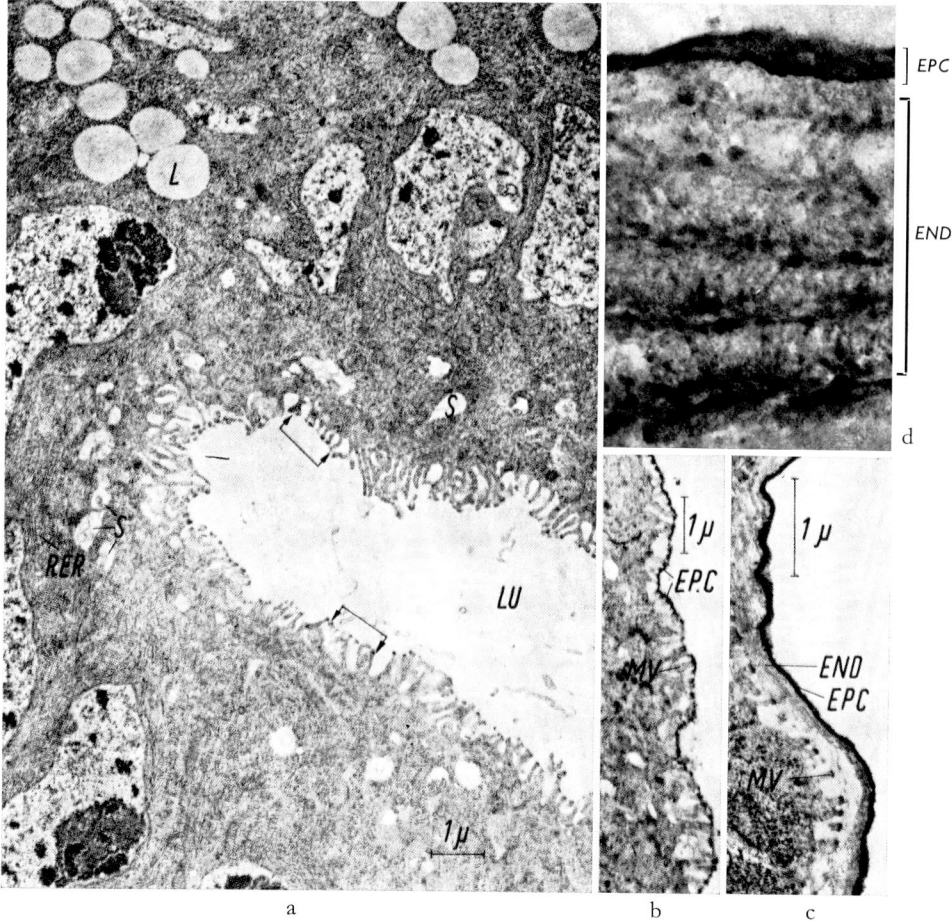

Fig. 5a—d. Changes at the microvillar surface of wing disks. a 124 h after oviposition; beginning of cuticle formation at tips of microvilli is seen between linked arrows. b 126 h; a continuous layer of epicuticle *(EPC)* is present. c 128 h; less dense, laminated material *(END)* lies between *EPC* and microvilli. d epicuticle and endocitucle as cast off by organism prior to 140 h ($\times 35000$). L lipid; LU lumen; RER rough endoplasmic reticulum; S "space". (From WEHMANN, 1969 and 1969a)

WEHMAN (1969, wing disk) did not see this extra layer. In accordance with earlier descriptions by BODENSTEIN (1950), he calls the entire cuticle laid down between puparium formation and pupation the prepupal cuticle. The cuticle that is produced later is then referred to as the imaginal cuticle by WEHMAN. Using this nomenclature, there is no true pupal cuticle. Both POODRY and SCHNEIDERMAN (1970) and WEHMAN (1969) agree that some time between pupation and 140 h, the newly formed cuticle is shed and begins to be replaced by the final cuticle. They also agree that by this time, the hypodermis is within three layers of cuticle: the terminal cuticle, the cuticle formed between the time of puparium formation and pupation, and the puparium. I feel the middle layer should be called pupal cuticle. The term prepupal cuticle then is no longer needed. Accordingly, the cuticle seen in the inside of metamorphosed, non-everted disk implants should be called pupal cuticle.

Summing-up these investigations thus far, it appears straightforward to recognize developmental stages of cells on the basis of their ultrastructural appearance. This becomes even more feasible at later stages of development. PERRY (1968) has carried out an extensive fine-structure analysis of the developing eye disk. Clearly, retinular cells can easily be recognized on the basis of such characteristics as the arrangement of microvilli, by the transformation of microvilli into rhabdomeres. Cone cells and pigment cells, too, are characterized ultrastructurally. Much more difficult, if not impossible, is the distinction between differentiating leg and wing cells. POODRY and SCHNEIDERMAN (1970) and WEHMAN (1969) describe the ultrastructure of cuticle-forming cells of leg and wing disks, respectively, during pupal stages; it is apparent from their electron micrographs that these cells look alike.

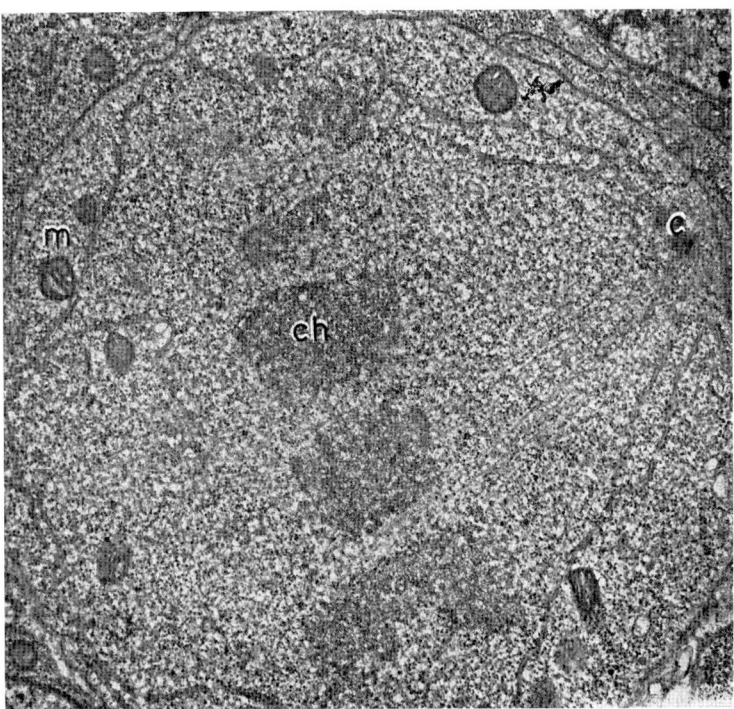

Fig. 6. Dividing cell in atelotypic (see GEHRING, this volume) culture of the 124th transfer generation, kindly provided by Prof. E. HADORN. *c* centriole; *ch* chromosome; *m* mitochondrion. Note spindle fibers reaching from centriole to chromosomes. Approx. ×20000. Unpublished electron micrograph by SCHABTACH

Nevertheless the data show that cells of early pupal eye disks can be told apart from leg and wing disk cells by their ultra-structure. Unpublished observations in our own laboratory showed that this holds also for cells of early pupal genital disks. Thus electron microscopy may prove useful for recognizing cell types not in larval, but in early pupal disks.

At later stages yet, when differentiation is on its way, dramatic fine structural differences are seen in the derivatives of the various disks. It is beyond the scope of this review to describe these, and the reader is referred to papers by SHOUP (1966), OVERTON (1967), PERRY (1968), WEHMAN (1969) and POODRY and SCHNEIDERMAN (1970).

Surprisingly few dividing cells have been seen in all these investigations (Figs. 2 and 6). Those that were seen in division were spherical, and maintained contact with the microvillar surface. Dividing cells have been seen in larval disks only in these studies, but the incidence is so low that one must not assign these isolated observations much significance.

B. Adepithelial Cells

Much less is known on fine structural changes during the development of adepithelial cells. In leg disks, where they have been studied in some detail by POODRY and SCHNEIDERMAN (1970) these cells were seen to line up tail-to-head, eventually forming long columns of cells filled with microtubules, and ultimately forming adult muscle. It must be stressed however that immigration of cells from elsewhere in the organism cannot easily be excluded in this *in vivo* study. Consequently, the evidence for the lineage from adepithelial cells to muscle is still circumstantial.

IV. Disk Eversion

Ontogenetically, disks are derived from local infoldings of the hypodermis. In the larval stage, their hypodermal origin is still manifested by a stalk that connects the disk to the hypodermis. The canal that must lead from the lumen of the disk to the outside of the animal has never been reconstructed in its entirety to my knowledge.

Eversion of the disk epithelium begins at the time of puparium formation and is completed some 12 h later; this point on the time axis is called pupation. The disk proper grows into the peripodial space, probably by the shape-transition from columnar to cuboidal of its epithelial cells, in a direction away from or at a right angle to the stalk. The lumen of the stalk widens, and the disk becomes exposed to the outside of the animal, with its microvillar surface facing out. Thus, relative surface orientations are not changed during the process of eversion. Parts of the peripodial membrane break down and are replaced by epithelial cells. Blood cells have been seen in the area of presumed peripodial cell death, perhaps phagocytizing degenerated cells of the stalk and/or peripodial membrane (POODRY and SCHNEIDERMAN, 1970).

V. Cell to Cell Communication and Adhesion

The electron micrographs of AGRELL (1966, 1968) show a network of secondary pores in the membranes of larval imaginal disk cells. These pores, it was thought, might serve the purpose of cell-to-cell communication. More recent investigations have shown however that the pores as they appear in AGRELL's pictures probably are artifacts produced by permanganate fixation (URSPRUNG and SCHABTACH, 1972). After glutaraldehyde fixation, these membranes appear as long, uninterrupted profiles and are certainly closer to the native configuration than those seen after permanganate fixation.

In their careful analysis of leg-disk fine structure, POODRY and SCHNEIDERMAN (1970) discovered cytoplasmic bridges connecting pairs of cells (Fig. 7). The bridges look like remnants of cytokinesis, and their lifespan is not known. The same structures were also seen in wing-disk electron micrographs by AGRELL (1968). It is not known whether more than two cells are ever interconnected by such bridges. Thus it is difficult at the moment to speculate meaningfully on their significance.

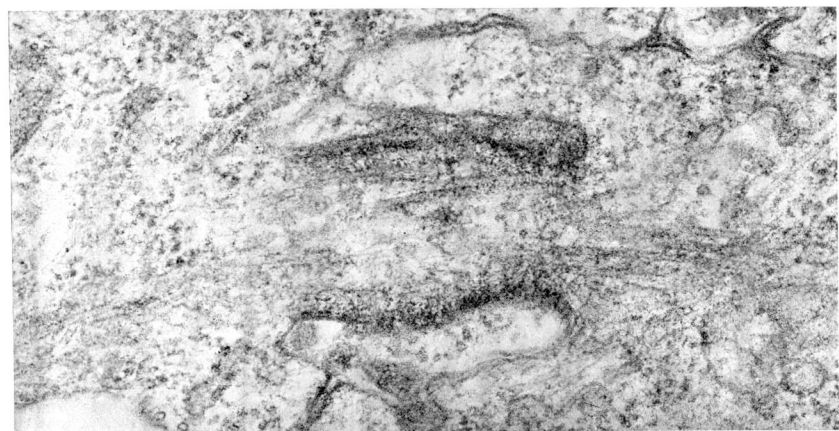

Fig. 7. Cytoplasmic bridge linking two leg-disk cells. The bridge contains microtubules and a dark, cylindrical border. × 59900. (From POODRY and SCHNEIDERMAN, 1970)

As is typical for epithelia, the cells of the epithelial portion of imaginal disks are interconnected by junctional complexes (Fig. 8). Starting at the microvillar surface, a zonula adhaerens leads to a region with septate desmosome structure, followed by zones variably described as gap junctions (POODRY and SCHNEIDERMAN, 1970), long fasciae adhaerentes (URSPRUNG and SCHABTACH, 1968) and occluding junctions (WEHMAN, 1969). The desmosome-like structures reported by URSPRUNG and SCHABTACH (1968) upon improved fixation and higher resolution are clearly septate as reported in the subsequent studies. POODRY and SCHNEIDERMAN (1970) purposely fixed some disks in hypertonic fixatives; after osmotic shrinkage, the cells were still attached to each other via gap junctions and septate desmosomes.

One might expect that frequency and perhaps location of junctional complexes in resting epithelia would differ from those in epithelia undergoing morphogenetic movements. With this question in mind, WEHMAN (1969) found that submicrovillar junctions are more complex in prepupae 128 h after oviposition than in younger disks, e.g., disks at the time of puparium formation. That is to say, junctional specializations in the case of wing disk development are more complex once the outgrowth of the disk is complete, which is the case of 128 h after oviposition.

POODRY and SCHNEIDERMAN (1970) also noticed changes in frequency of junctional complexes as a function of developmental stage. Specifically they report that adepithelial cells display gap junctions in pupal, but not in larval cells. The epithelial cells in their investigation on the other hand were seen to contain all varieties of junctional specializations in larval and pupal disks.

I feel these results should be considered preliminary because of the severe handicap posed by sampling error in thin sections as they are used in electron microscopy. Also, in view of the dynamic model of junction formation presented by CAMPBELL and CAMPBELL (1971) it is perhaps erroneous to expect simplification or disappearance of adhesive specializations during morphogenetic movements.

VI. Fine Structure of Mutant Imaginal Disks

A. Cell Death

Cell death is known to play an important role in morphogenesis (see SAUNDERS, 1966, for a review). In the case of imaginal disks, a study of cell death promised not only insight into normal morphogenesis, but was a logical approach to the explanation of mutants that lack portions of adult appendages.

With the question in mind, FRISTROM (1968, 1969) studied the fine structure of wild type eye and wing disks, of the wing-mutants *miniature*, *vestigial*, *apterousXasta*, *Beadex*, and *cut*, and of the eye-mutants *Bar*, *Double Bar*, *eyeless*, and *lozenge*. Clumped appearance of chromatin, exceedingly dense packing of ribosomes, "myeline" structures and fragmentation of cells, perhaps coupled with fine-structural irregularities of membranes, were taken as criteria for cell degeneration (Fig. 9). Also, some instances of apparent phagocytosis were seen. The correlation of these ultrastructural peculiarities with cell degeneration is made very likely by lightmicroscope analysis of adjacent sections; in these sections, some cells were contrasted with toluidine blue, which stains degenerating bodies dark.

The frequency of degenerating cells was seen to differ in the various mutants. No such cells were observed in wild type or *miniature* disks at any stage investigated, whereas they were seen frequently in early third instar wing disk of *apterous* larvae, and late third instar *vestigial* and *Beadex*

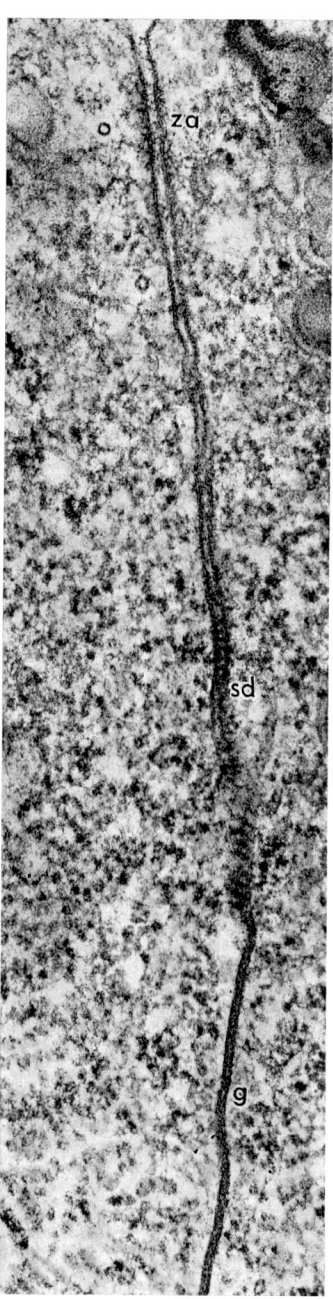

Fig. 8. Succession of junctional specializations between epithelial cells of larval leg disks. Note zonula adhaerens *(za)* near lumen and then, towards the basal regions of the cell, a septate desmosome *(sd)* and an extended gap junction *(g)*. ×67000 (From POODRY and SCHNEIDERMAN, 1970)

disks, e.g. Wild-type eye disks contain some moribund cells, but *Bar* and *eyeless* disks contain more.

In a recent light-microscope study, SPREIJ (1971) reports rather extensive cell death to occur in wild type wing disks, where it was not seen in the electron microscope studies by FRISTROM (1969), nor by WEHMAN (1969). The incidence of cell

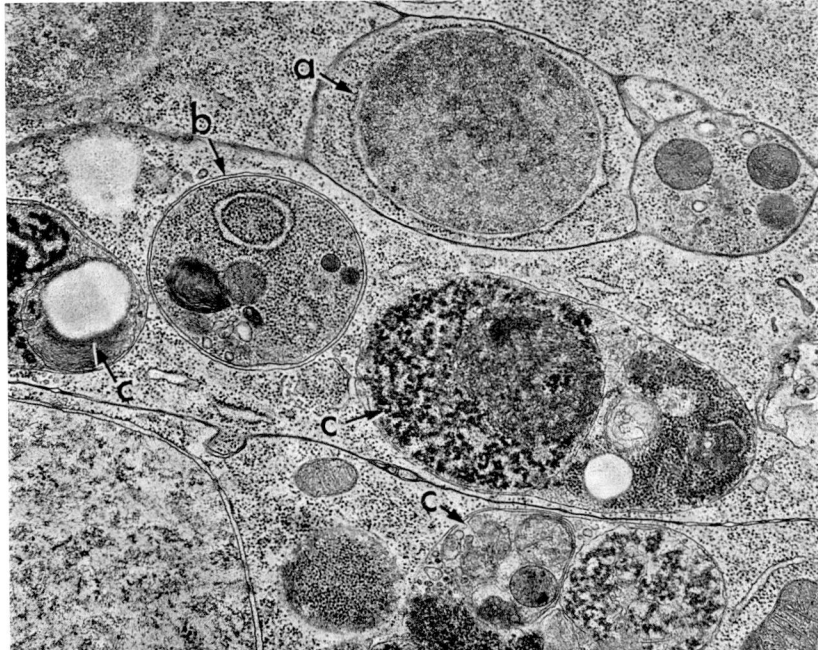

Fig. 9. Cell fragments from a *Bar* eye disk in various stages of degeneration. *a* condensed nuclear fragment not yet phagocytosed; *b* cytoplasmic fragment engulfed but still surrounded by two membranes; *c* fragments bounded by single membranes and showing cytolysis. × 17000 (From FRISTROM, 1969)

death will have to be kept in mind in interpreting asymmetries of spot size that are observed in cell lineage experiments involving somatic recombination (see NÖTHIGER, this volume).

B. "Neoplasms"

A recent new mutant, *lethal giant larva* $(l(2)gl^4)$ upon electron microscope examination was seen to contain highly abnormal imaginal disks (GATEFF and SCHNEIDERMAN, 1969). In gross morphology, the mutant disks differ from the normal in shape, and are exceedingly large. At the fine structural level, the disks are seen to have been invaded by blood cells; the disk cells proper have largely lost their mutual adhesion, lack junctional specializations, and exhibit abnormal shape. They form an amorphous mass of tissue rather than an orderly folded epithelium. Upon transplantation into wild-type hosts, the mutant tissue grows extensively into amorphic tissue masses and invariably kills the host; hence, its classification as a neoplasm.

VII. Non-autonomous Development of Disks: Ultrastructural Evidence for Uptake of Materials from Larval Hemolymph

For studies of the molecular differentiation of cells and tissues it is important to know which syntheses are carried out autonomously by cells; the mere recording of the presence of a macromolecule in a cell does not constitute proof that the molecule was synthesized by this cell.

POODRY and SCHNEIDERMAN (1970) traced horseradish peroxidase injected into the hemolymph of larvae and found that this marker protein is taken up into imaginal disk cells within two hours after injection. It (and one would assume, other proteins too) appears to be taken up by pinocytosis.

VIII. Virus-Like Particles

AKAI et al. (1967) show electron micrographs of rapidly dividing *Drosophila* cells derived from imaginal disks; these cells contain numerous virus-like particles, some 377×455 Å big, slightly elliptical. The same authors also found these particles in larval brain tumors, blood and gut cells. In the latter, similar if not identical particles were also seen by FILSHIE et al. (1967, 1971), particularly in the interstitial and cuprophilic cells. The same authors (1967) had also observed such particles in cultured *Aedes aegypti* cells. In the prothoracic ganglion of normal ants, similar particles have been reported by STEIGER et al. (1969). Other *Drosophila* cell types in which these particles were seen include adult midgut, treacheae, paragonia, and nurse cells (RAE and GREEN, 1968), adult fat body, oenocytes, and brain (PHILPOTT et al., 1969), and melanotic masses (PEROTTI and BAIRATI, 1968). WEHMAN (1969) found such structures in nuclei of cells of wing-disks that had been held in hanging-drop cultures. His recent study (WEHMAN and BRAGER, 1971) tentatively classifies the particles as virions of a nonenveloped, capsid virus with capsomers of 55 Å diameter, similar to polyoma. In our laboratory, we have never observed such particles in normal wing or genital disks, but have encountered them in large numbers, sometimes in cristalline arrangement, in both nuclei and cytoplasm of disk-derived cells kindly supplied by Prof. E. HADORN (unpublished). Among those, virus-like particles were particularly numerous in cell lines that are characterized by their rapid proliferation. This observation confirms that of AKAI et al. (1967).

It is interesting that particles of similar size and shape have also been seen in cultured plant cells, in a report by SJOLUND and SHIH (1970). These authors showed such particles in a cotyledon-derived cell-line of the Cruciferan *Streptanthus tortuosus var. orticulatus*. This is a cell-line that according to the authors has lost the ability to differentiate, much like the so-called atelotypic *Drosophila* cell-lines of HADORN (1966).

IX. Conclusions

1) Four cell types have been found to build up larval imaginal disks: the epithelial cells of the disk proper, which are arranged in a folded, essentially single-layered epithelium, and which, from their microvillar surface, form the cuticular structures; the adepithelial cells, which perhaps give rise to musculature; nerve cells; tracheolar cells. The disk as a whole is enclosed in a basement lamina. The so-called peripodial membrane is continuous with the folded epithelium.

2) No firm evidence is available that would enable one to distinguish epithelial cells of different larval disks, or different organ-forming areas within larval disks, from one another at the fine structural level, except that retinular cells of mature eye disks occur in groups of eight and are spindle-shaped.

3) Larval disks cells do not molt. At the time of puparium formation, the pupal cuticle begins to be laid down on the microvillar surfaces of the disks. Several other finestructural changes occur from then on, such as an increase in the incidence of rough ER.

4) Surface relationships do not change during disk eversion; the microvillar surface remains the "outside". The surface-increase that accompanies eversion results largely from the transition from columnar to cuboidal of the epithelial cells.

5) Several kinds of junctional specializations are present in imaginal disks, including cytoplasmic bridges. Their possible role in cell-cell communication is unknown.

6) Local cell death has been seen in several wild-type disk epithelia, and in the peripodial membrane during disk eversion.

7) Disk cells take up horse-radish peroxidase from hemolymph.

8) Morphological mutants that result in the loss of derivatives of imaginal disks are reflected, ultrastructurally, by abnormally high frequency of cell death. A mutation that removes normal control of disk growth is reflected by chaotic arrangement of disk cells and absence of junctional specializations. Virus-like particles have occasionally been seen in normal disks, and are abundant in disks grown in organ culture both *in vivo* and *in vitro*. Atelotypic cell-lines invariably contain such particles.

Acknowledgements

Critical comments on this manuscript by Drs. D. FRISTROM, H. A. SCHNEIDERMAN, TH. E. SPREIJ, and H. J. WEHMAN are gratefully acknowledged.

References

AGRELL, I. P. J.: Continuity of the membrane systems in the cells of imaginal disks. Z. Zellforsch. **72**, 22–29 (1966).

AGRELL, I. P. S.: Differentiation of the membrane systems in the cells of imaginal disks. Z. Zellforsch. **88**, 365–369 (1968).

AKAI, H., GATEFF, E., SCHNEIDERMAN, H. A.: Virus-like particles in normal and tumorous tissues of *Drosophila*. Science **157**, 810–813 (1967).

AKAI, H., GATEFF, E., DAVIS, L. E., SCHNEIDERMAN, H. A.: Organization and fine structures of wild-type and cultured eye-antennal imaginal disks of *Drosophila*. 1969 (Quoted in GATEFF and SCHNEIDERMAN, 1969).

CAMPBELL, R. D., CAMPBELL, J. H.: Origin and continuity of desmosomes. In: REINERT, J., URSPRUNG, H. (Eds.): Origin and Continuity of Cell Organelles. Vol. 2 of Results and Problems in Cell Differentiation. Berlin-Heidelberg-New York: Springer pp. 261–298 1971.

CHIARODO, A. J., DENYS, F. R.: Fine structural features of developing leg disks of the blowfly, *Sarcophaga bullata*. J. Morph. **126**, 349–364 (1968).

FILSHIE, B. K., GRACE, T. D. C., POULSON, D. F., REHACEK, J.: Virus-like particles in insect cells of three types. J. Invert. Pathol. **9**, 271–273 (1967).

FILSHIE, B. K., POULSON, D. F., WATERHOUSE, D. F.: Ultrastructure of the copper-accumulation region of the *Drosophila* larval midgut. Tissue and Cell **3**, 77–102 (1971).

FRISTROM, D.: Cellular degeneration in wing development of the mutant vestigial of *Drosophila melanogaster*. J. Cell. Biol. **39**, 488–491 (1968).

Fristrom, D.: Cellular degeneration in the production of some mutant phenotypes in *Drosophila melanogaster*. Molec. Gen. Genetics **103**, 363—379 (1969).

Gateff, E., Schneiderman, H. A.: Neoplasms in mutant and cultured wild-type tissues of *Drosophila*. National Cancer Institute Monograph **31**, 365—397 (1969).

Overton, J.: The fine structure of developing bristles in wildtype and mutant *Drosophila melanogaster*. J. Morph. **122**, 367—380 (1967).

Perotti, M. E., Bairati, A. J., Jr.: Ultrastructure of the melanotic masses in two tumorous strains of *Drosophila melanogaster* (tu B_3 and Freckled). J. Invertebr. Pathol. **10**, 122—138 (1968).

Perry, M. M.: Further studies on the development of the eye of *Drosophila melanogaster*. 1. The ommatidia. 2. The interommatidial bristles. J. Morph. **124**, 227—261 (1968).

Philpott, D. E., Weibel, J., Altan, H., Miquel, J.: Viruslike particles in the fatbody, oenocytes and central nervous tissue of *Drosophila melanogaster* imagoes. J. Invertebr. Pathol. **14**, 31—38 (1969).

Poodry, C. A., Schneiderman, H. A.: The ultrastructure of the developing leg of *Drosophila melanogaster*. Wilhelm Roux' Arch. Entwickl.-Mech. Org. **166**, 1—44 (1970).

Rae, P. M. M., Green, M. M.: Viruslike particles in adult *Drosophila melanogaster*. Virology **34**, 187—189 (1967).

Saunders, J. W.: Death in embryonic systems. Science **154**, 605—612 (1966).

Shoup, J. R.: The development of pigment granules in the eye of wild type and mutant *Drosophila melanogaster*. J. Cell. Biol. **29**, 223—249 (1966).

Sjolund, R. D., Shih, C. Y.: Viruslike particles in nuclei of cultured plant cells which have lost the ability to differentiate. Proc. nat. Acad. Sci. (Wash.) **66**, 25—31 (1970).

Spreij, Th. E.: Cell death during the development of the imaginal disks of *Calliphora erythrocephala*. Netherl. J. Zoology **21**, 221—264 (1971) (This article contains information on cell death in *Drosophila* disks, too.).

Steiger, U., Lamporter, H. E., Sandri, C., Akert, K.: Virus-ähnliche Partikel im Zytoplasma von Nerven- und Gliazellen der Waldameise. Arch. ges. Virusforsch. **26**, 271—282 (1969).

Ursprung, H., Schabtach, E.: The fine structure of the male *Drosophila* genital disk during late larval and early pupal development. Wilhelm Roux' Arch. Entwickl.-Mech. Org. **160**, 243—254 (1968).

Ursprung, H., Schabtach, E.: On the syncytial nature of *Drosophila* imaginal disks: Influence of fixatives on membrane fine structure. Rev. suisse Zool. In Press (1972).

Waddington, C. H., Perry, M. M.: The ultrastructure of the developing eye of *Drosophila*. Proc. roy. Soc. B **153**, 155—178 (1960).

Wehman, H. J.: Fine structure of *Drosophila* wing imaginal disks during early stages of metamorphosis. Wilhelm Roux' Arch. Entwickl.-Mech. Org. **163**, 375—390 (1969).

Wehman, H. J.: Fine structure of the developing wing imaginal disk in *Drosophila*. Dissertation. The Johns Hopkins University, Baltimore (1969a).

Wehman, H. J., Brager, M.: Viruslike particles in *Drosophila*: Constant appearance in imaginal disks *in vitro*. J. Invertebr. Pathol. **18**, 127—130 (1971).

The Biochemistry of Imaginal Disk Development[1]

James W. Fristrom

Department of Genetics, University of California, Berkeley, California

I. Introduction

The extensive developmental studies conducted by Ernst Hadorn, his students and his colleagues, and reviewed in this volume, have made imaginal disks of *Drosophila* one of the most extensively characterized embryonic tissues known. Through numerous investigations on determination, transdetermination and differentiation imaginal disks have emerged as not only convenient tissues for biological studies, but also for molecular studies. Many important requirements for molecular biological investigations on development are met by disks. First, *Drosophila melanogaster* is by far the best genetically characterized multicellular organism and therefore the important advantages supplied by genetic techniques are available. Furthermore, most of the known mutants of *Drosophila* affect disks or structures derived from disks. Thus, mutants are available which affect viability of specific cells within disks (Fristrom, 1969a; Spreij, 1971), the viability of all disks (El Shatoury and Waddington, 1957; Stewart et al., 1972), the developmental fate of disks (Lewis, 1964; Gehring, 1966; Ouweneel, 1969a, b), growth of disks (Gateff and Schneiderman, 1969; Stewart et al., 1972), and possibly the capacity of disks to differentiate at metamorphosis (Stewart et al., 1972). Additionally the systematic recovery and use of temperature sensitive mutants (Suzuki, 1970) offers an important tool both to the developmental and the molecular biologist. The characteristics of imaginal disk differentiation during the prepupal period (which we call "disk metamorphosis" in this monograph) are also noteworthy. Disk metamorphosis is controlled by insect molting hormone (ecdysone) and juvenile hormone (see the article by Oberlander in this volume and the discussion below). In addition, responses of disks to both of these hormones occur in culture (Oberlander and Fulco, 1967; Agui et al., 1969; Sengel and Mandaron, 1969; Mandaron, 1970, 1971; Fristrom et al., 1969; Chihara et al., 1972) and thus the unique opportunity for studying disk metamorphosis *in vitro* is available. From the biochemical viewpoint the limitations resulting from the small size of disks can in part be overcome by the use of mass isolation procedures (Fristrom and Mitchell, 1964; Fristrom and Heinze, 1968; Zweidler and Cohen, 1971). Early metamorphosis of disks involves not only changes in disk morphology, but also the secretion and deposition of the cuticle.

---

[1] Previously unpublished material presented in this monograph was supported in part by grants from the National Science Foundation (GB-8176, GB-29290) and under contract with the Atomic Energy Commission (AT(O4-3)-34).

Since cuticles are composed of large quantities of proteins and chitin the production of cuticle presumably involves a substantial synthetic output by the disks and offers a convenient handle for the study of the regulation of protein synthesis and enzyme function. Taking the above discussion into consideration it is clear that imaginal disks are particularly well suited for the study of the modes of action of insect hormones. In this context, it must be emphasized that a discussion of the biochemistry of imaginal disk development cannot be made without reference to the action of insect hormones upon disks.

II. Methodology

A. The Preparative Isolation of Imaginal Disks

The first technique for the preparative isolation of imaginal disks (FRISTROM and MITCHELL, 1965) yielded 2000—4000 disks. A modification (FRISTROM and HEINZE, 1968) produced disk preparations containing approximately 40 000 disks which were 90—93% free of other tissue debris. The isolation procedure has been modified again and now produces an average yield of 220 000 disks per day and an average purity of 96% (based on yields from 24 preparations). The new isolation procedure takes about 5 h. The details are as follows:

Approximately 500 gm of larvae are ground in a Universal laboratory grinding mill (obtained through VWR Scientific: the same mill as used previously, but now

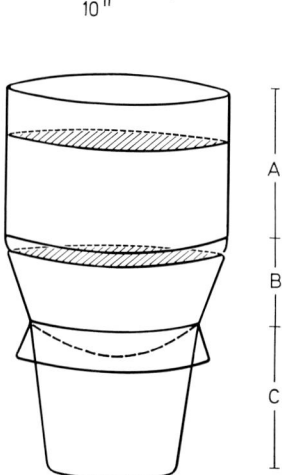

Fig. 1. Sieving apparatus for preparative isolation of imaginal disks. The screens and the different sections (designated A, B, C) are all separable and constructed of stainless steel

sold under a different name) in 400 ml of cold (4° C) RINGER's solution (EPHRUSSI and BEADLE, 1936) directly into the receptacle pictured in Fig. 1. The first screen has 2 mm openings, the second, 0.8 mm openings. A piece of silk bolting cloth (size 8 xx) is fitted in the shape of a shallow bag over a 13 qt. stainless steel bucket. The bucket is surrounded with ice. The two screens and cloth are washed in succession with a gentle spray of Ringer's driven from a pressurized reservoir at 12 p.s.i. The final volume in the bucket is approximately 8 l.

The material in the bucket is allowed to sediment for 14 min and the volume is then reduced to 1400 ml by aspirating fluid from the surface. The material is then transferred to a 1500 ml beaker, allowed to sediment for 8 min and the volume is reduced by aspiration to 200 ml. Ringer's is added to bring the volume to 600 ml, the material allowed to sediment for 6 min, and the volume is then reduced to 200 ml. After bringing the volume to 400 ml, and allowing the tissue to settle for 5 min, the volume is now reduced to 100 ml. This last step is repeated 2 or 3 more times until the supernatant is clear at the end of the 5 min settling period.

The disks are suspended in 200 ml of Ringer's and equally distributed into approximately 20 clean, dry Petri dishes (100 mm). The disks adhere to the glass more firmly than the contaminating tissue. When the fluid is poured off, after standing for 4 min, the disks tend to remain attached to the glass and are then immediately washed from the Petri dish with a spray of Ringer's solution. The decanted material is poured into a second set of Petri dishes and the disks recovered are pooled with those obtained from the first set of dishes.

The disks are next purified with the use of a discontinuous Ficoll gradient. The material is pelleted at 900 RPM in a clinical centrifuge and then taken up in 10 ml of 14% Ficoll made up in Ringer's solution. The material is layered on top of 2 gradients containing a bottom layer of 25 ml of 21% Ficoll and a top layer of 35 ml of 14% Ficoll contained in 45 × 125 mm polycarbonate centrifuge tubes. The gradients are spun at 800 RPM for 6 min in a Sorvall GLC centrifuge in an HL-4 rotor. The disks are recovered from the gradient interface with a prewetted Pasteur pipet. The Ficoll is then diluted out by washing the disks several times in Ringer's.

The disks are now poured through a series of Petri dishes as depicted in Fig. 2 and are eventually accumulated in glass beakers. They are then given a final cleaning by removing pieces of tissue debris with a fine glass needle attached to an aspirator. They can be counted by distributing them in a Petri dish placed over a sheet of millimeter graph paper or be weighed. To determine wet weight the disks are pelleted at 900 RPM in a clinical centrifuge, the Ringer's is decanted and the tube drained and then weighed, and the weight of the disks determined by subtracting the weight of the empty tube which had been exposed to the same manipulations. A preparation of 200000 disks weighs about 0.4—0.45 gm. When a correction is made for fluid associated with the disk pellet, the wet weight of 100000 disks is 0.1—0.15 gm. The distribution of disk types is as follows: Wing: 18%, Leg: 55%, haltere: 7%, genital: 2%, eye: 14%, antennal: 4%. The labial and clypeo-labrum disks which give rise to head structures and intact eye-antennal complexes are not recovered in appreciable quantities. A preparation of disks obtained by using the above procedure is pictured in Fig. 3.

An elegant and elaborate procedure for the isolation of organs of *Drosophila melanogaster* has recently been developed by Zweidler and Cohen (1971). This procedure has a distinct advantage over the one described above in that it allows the recovery of other organs (particularly salivary glands). The yield of disks from 500 gm of larvae is about 0.4 gm wet weight. It is not clear if this value has been corrected for disk associated fluid. However, the technique at least gives yields of

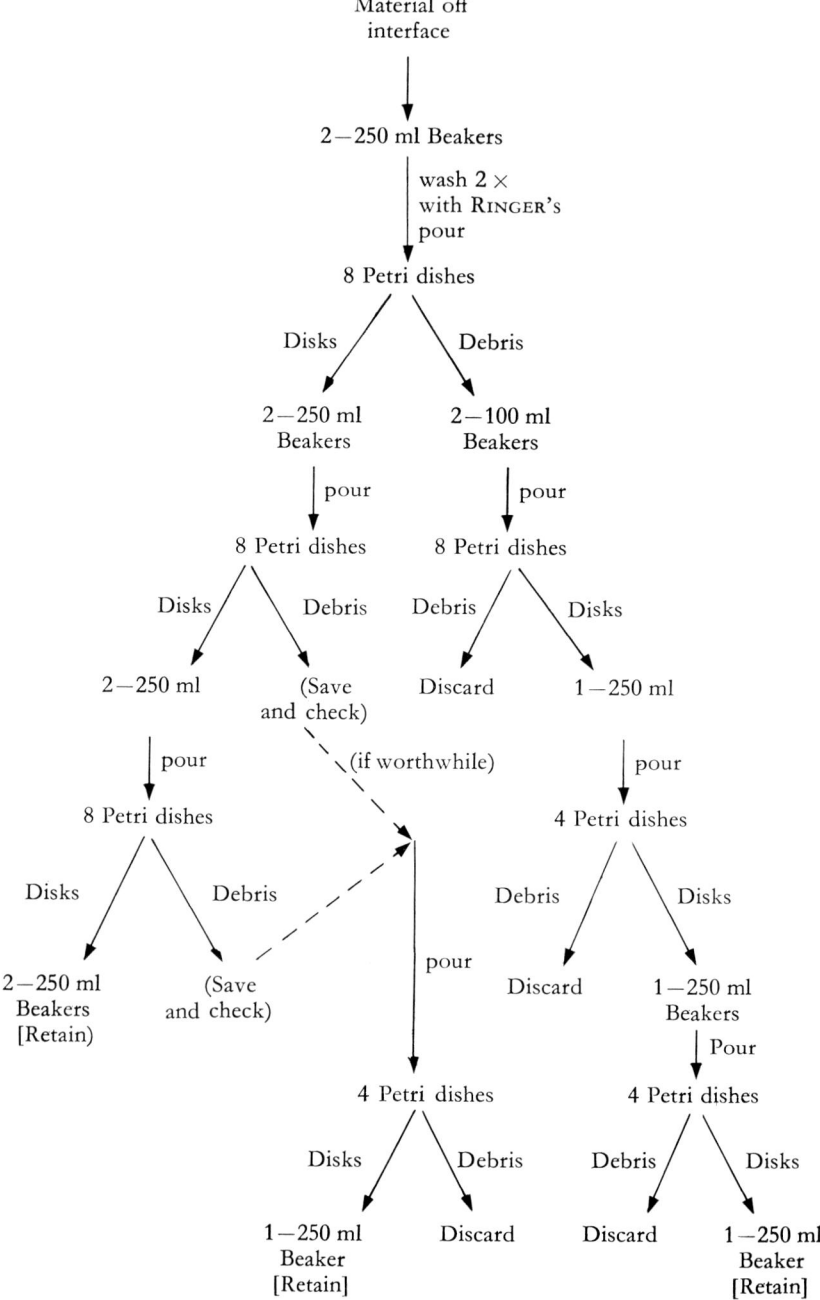

Fig. 2. Purification of imaginal disks by attachment to a clean glass surface. Disks are allowed to settle in each dish approximately 4 min before the fluid is decanted. The disks tend to stay in the dishes attached to the glass surface while other tissues are generally decanted with the RINGER'S

disks which are as great as those obtained in our laboratory. However, the purity of the disks, described by the authors as being 80% on a tissue mass basis, appears to be lower than the purity achieved by the above procedure.

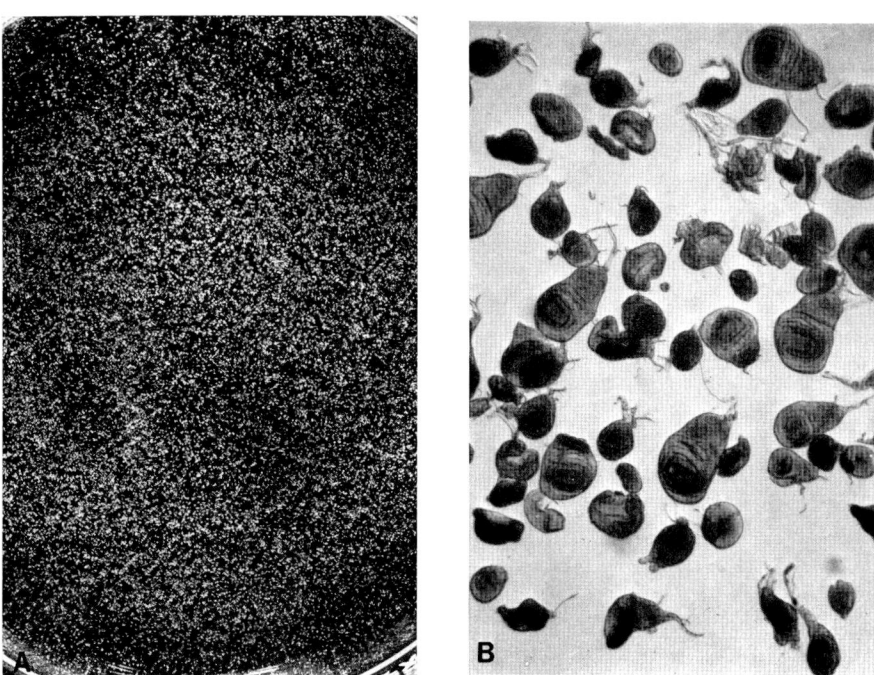

Fig. 3. A Preparative disk isolate in a 100 mm Petri dish. The dish contains an estimated 100000 disks (1 ×). B A sample of the disk population in the dish shown at high power (40 ×) (Photographs by PETRI and CHIHARA)

B. Culture Conditions

Disks of *D. melanogaster* are capable of normal development *in vitro*. SCHNEIDER (1964) developed a culture medium in which eyeantennal disks cultured with other larval tissues (in particular the brain) are capable of differentiation. Using a modification of SCHNEIDER's medium, MANDARON (1970) has demonstrated that a variety of disk types cultured with the larval brain and with α-ecdysone are capable of complete differentiation. In a recent paper, MANDARON (1971) reported the development of a completely defined medium in which evagination and differentiation of leg and wing disks cultured with α-ecdysone occurs. The importance of this achievement for the future study of imaginal disk development is immense. It also demonstrates unequivocally that the *in vitro* culture conditions maintain the disks in a biologically normal state.

Another defined medium used for the culture of *Drosophila* imaginal disks was developed by ROBB (1969). Evagination of leg and wing disks also occurs in this medium in the presence of ecdysones (FRISTROM et al., 1969; CHIHARA et al., 1972). Of importance for biochemical studies is the fact that evagination occurs in mass culture (5000 disks/ml) with a fair amount of synchrony (Fig. 4). However, complete disk differentiation in Robb's medium has not been reported. The maintenance of biological integrity of disks cultured in Robb's medium is demonstrated by the fact

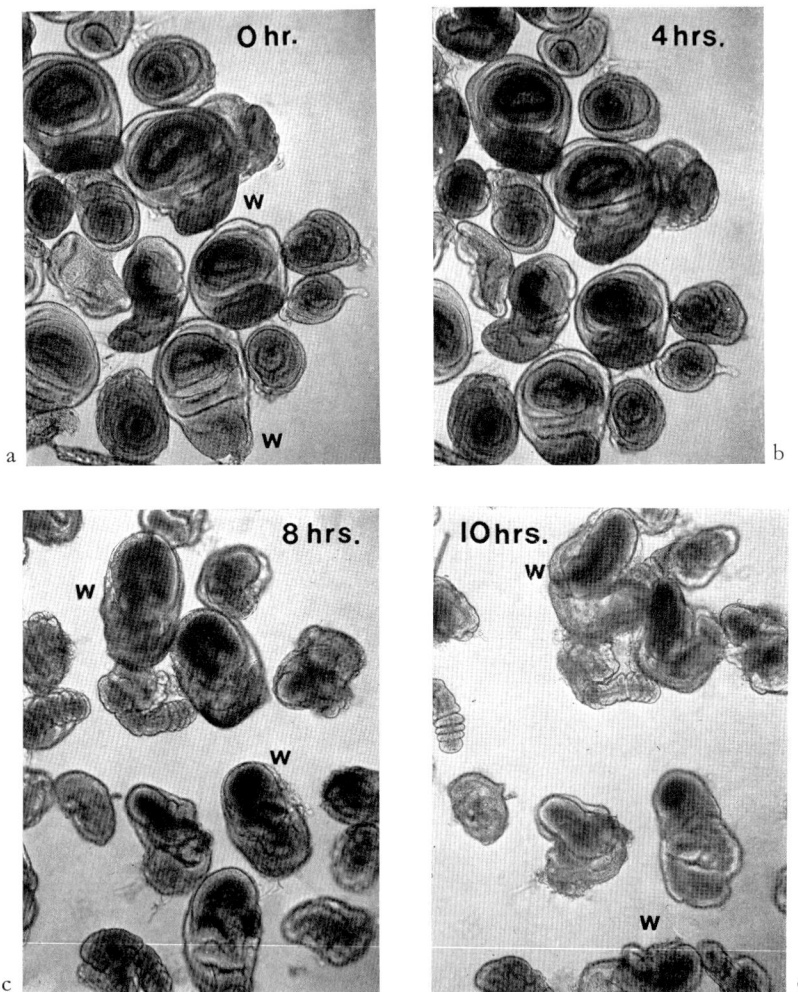

Fig. 4a—d. Evagination of mass isolated disks. Note the general synchrony of evagination in the different disks and particularly in 4 wing disks *(w)*. Full evagination is obtained after 12—14 h and is not shown in these photographs (60 ×) (Photographs by Petri and Chihara)

that after 74 h of culture, wing disks are still capable of differentiation when transplanted to developing larvae (Robb, 1969). As fas as synthesis of macromolecules is concerned, there are no obvious differences in the rates of incorporation into RNA and protein in Robb's and Schneider's media during early phases of disk culture (Fristrom, 1970).

III. Biochemical Composition of Imaginal Disks

A. DNA

Disk cells are generally believed to be predominantly diploid. Estimates of the number of cells in disks have been provided by several workers using cell-lineage experiments. The number of cells in a late third instar wing disk is estimated to be approximately 16000—17000 (BRYANT, 1970). GARCIA-BELLIDO and MERRIAM (1971) find about 52000 cells present in wing disk tissue 18 h after puparium formation. Since cell divisions are infrequent following puparium formation, then late larval disks would be expected to contain 25000—30000 cells. Eye disks of late third instar larvae contain about 30000 cells (BECKER, 1957). Data presented by BRYANT and SCHNEIDERMAN (1969) allow an estimation of the number of cells in a late third instar leg disk of about 2000 cells (based on 20 prospective leg cells with a division time of 15 h). By comparisons of relative sizes of disks this estimate seems low and suggests that the estimated division time may be somewhat high. Estimates of the diploid DNA content of *D. melanogaster* cells range from 0.14—0.40 pg/nucleus (KURNICK and HERZKOWITZ, 1952; TARTOF and PERRY, 1970) with an intermediate value of 0.28 pg/nucleus being preferred by LAIRD (1971) because of its agreement with DNA/DNA hybridization data. Based on these estimates, late larval wing disks should contain about 4.2 ng of DNA; eye-antenna disks about 8.4 ng of DNA. Direct estimates of DNA content of preparatively isolated disks by STEWART (1971) using the procedure of MUNRO and FLICK (1966) indicate that the average disk from a third instar larva contains 6 ng of DNA and therefore about 21500 cells (using 0.28 pg/diploid nucleus). Since mass-isolated disk preparations predominantly contain leg disks this estimate is surprisingly high. One possible source of error, contamination by polytene tissues, was investigated. It was found, however, that the average fragment of tissue debris contained somewhat less DNA than the average disk (unpublished observation), and therefore the tissue debris makes no appreciable contribution to the total amount of DNA recovered. Furthermore, the average larval wing disk was found to contain about 15 ng of DNA and therefore contain about 53000 cells. This estimate is larger than those based on cell lineage studies (16000—17000 or 25000—30000). Estimates based on cell lineage studies are for many reasons underestimates of the number of cells present in larval disks. In particular, cell counts are not readily obtainable on all regions of the adult epidermis and some areas are not sufficiently marked to allow detection of all cells. Also there are cells present in disks (in the peripodial membrane and the adepithelium) which do not give rise to adult cuticular structures and therefore are not detected in cell lineage experiments. Finally, cells in the disk which are about to divide will have "tetraploid" quantities of DNA and estimates based on the DNA content of somatic nuclei will tend to overestimate the number of cells present. For all of these reasons the estimates of the number of cells in disks based on DNA content would not be expected to agree precisely with those produced by somatic genetic experiments and hence the estimates produced by the two methods are not unreasonably different. It is obvious, however, that the results from the two sources would be in even better agreement using the estimate of 0.4 pg/somatic nucleus found by TARTOF and PERRY (1970) rather than the value of 0.28 pg/nucleus.

Disk DNA is similar to that found in other non-polyene tissues (eg., nervous tissue, embryonic tissue) in that it contains readily demonstrable quantities of satellite DNA which is apparently localized in the centromeric heterochromatin (GALL et al., 1971). The number of rRNA genes present in disk DNA is similar to that found in DNA from adult tissues of *D. melanogaster* and is not amplified even in *bobbed* mutants which are deficient in rRNA genes (STEWART, 1971).

B. RNA

Disks stain heavily with basic dyes and have long been assumed to contain relatively large amounts of RNA. Electron micrographs of larval disk cells show that the cytoplasm is packed with free ribosomes (URSPRUNG and SCHABTACH, 1968; WEHMAN, 1969; POODRY and SCHNEIDERMAN, 1970). The ratio of RNA/DNA is not, however, dramatically high, being about 4.3:1 (Table 1), which is similar to the ratio found in liver cells. The low RNA/DNA ratio probably stems from the fact that disk cells are very small, and although the cytoplasm is packed with ribosomes, the nucleus occupies substantial space within the disk cell.

Table 1. Biochemical composition of imaginal disks

Wet weight	DNA	RNA	Protein	RNA/DNA	Protein/RNA
1.0–1.5 µg	0.006 µg	0.026 µg	0.135 µg	4.33	5.2

Values are in micrograms/disk.
Data are based on mass isolated disks and represent the average disk found in a preparation (STEWART, 1971).

C. Protein

The protein content per disk is about 135 pg for preparatively isolated disks (Table 1). SDS-acrylamide gel electrophoresis resolves disk proteins into numerous distinct species (Fig. 16), however, little is known about the types of proteins found in disks, or about their distribution within and between different types of disks. A very useful technique to locate proteins in disks has been developed by SPREY (1970) who has been able, using whole mount preparations and histochemical techniques, to localize some enzymes in disks. This technique (discussed below) provides a means for determining the distribution of many enzymes within disks, and therefore allows the development of "enzyme maps" of disks analogous to the fate maps prepared by developmental biologists.

IV. Physiology and Metabolism of Insect Hormones

A. Ecdysones

The *in vitro* response of imaginal disks to ecdysones offers the unique opportunity for studying the mechanism of action of the hormone. However, in addition to considerations about the action of ecdysone upon disks are considerations about the specific structure of the hormone which is responsible for *in situ* molting hormone activity, the uptake and distribution of ecdysone within the cells of the disks and the possibility of metabolism of the hormone by the disks.

The problem of determining the specific structure of ecdysone responsible for molting hormone activity is best approached by comparing concentrations of hormone required for *in vitro* responses with those required for *in situ* responses. Unfortunately, the determination of the physiological concentration for ecdysone activity is difficult and imprecise. In Table 2 some comparisons are made between ecdysone titres found during pupal molts and amounts of ecdysone required to induce assayable responses in insects after injection. As can be seen in the table, generally more hormone must be injected to induce a response than is found *in situ*. This condition probably results, in part, from the fact that ecdysones are rapidly metabolized by insects. Thus, in *Bombyx* larvae, MORIYAMA and his coworkers (MORIYAMA et al., 1970) demonstrated the rapid conversion of α- to β-ecdysone and the subsequent degradation of that molecule to inactive products. These authors found that the half-life of the hormone injected into 6 day old last instar larvae was very short with only about 5% of the injected α-ecdysone being recovered after 2 h. The conversion of β-ecdysone to degradation products was slower than the conversion of α-ecdysone to β-ecdysone. The degradation of β-ecdysone was particularly slow in mature larvae. The rapid metabolism of α-ecdysone to β-ecdysone and the degradation of the latter compound to inactive molecules is apparently a widespread condition in insects (KING, personal communication). It is therefore not surprising that excessive amounts of ecdysones must be injected in order to produce an assayable response *in vivo*.

Table 2. Estimations of physiological concentrations of ecdysones

Insect	Hormone	Amount injected µg/gm	*In situ* titre µg/gm	Equivalent molar concentration
Sarcophaga	α-ecdysone	0.640[b]	—	1.4×10^{-6} M
	β-ecdysone	0.330[b]	—	6.9×10^{-7} M
Musca	α-ecdysone	0.310[c]	—	6.7×10^{-7} M
	β-ecdysone	0.310[c]	—	6.5×10^{-7} M
Calliphora	α-ecdysone	0.500[c]	—	1.1×10^{-6} M
	β-ecdysone	0.470[c]	—	1.0×10^{-6} M
	[a]	—	0.124[d]	2.7×10^{-7} M
Bombyx	[a]	—	0.108[e]	2.3×10^{-7} M
	[a]	—	0.216[f]	4.6×10^{-7} M
Milkweed Bug	[a]	—	0.031[g]	6.7×10^{-8} M

[a] Form of ecdysone recovered is not specified. Molecular weight of α-ecdysone used to determine molar concentration.
[b] From OHTAKI, MILKMAN, and WILLIAMS (1967).
[c] From KAPLANIS et al. (1966).
[d] From SHAAYA and KARLSON (1965b).
[e] From BURDETTE (1962) (larval titre).
[f] From SHAAYA and KARLSON (1965a) (pupal titre).
[g] From FEIR and WINKLER (1969).

Because injected ecdysones are rapidly metabolized and degraded the level of the *in situ* titre of ecdysone is probably a better estimate of the physiological hormone concentration needed to produce a response than that provided from injection data.

In order to compare *in situ* titre data with concentrations of ecdysone in solution, the quantity of hormone per gram of insect has been equated to quantity per milliliter of solution and then expressed in terms of molar concentration (Table 2). This manipulation is intended to offer only an order of magnitude estimate of the physiological concentration needed to induce a response. As such, it might be required that

Table 3. Induction of evagination with ecdysones

Ecdysones	Concentration required	
	Half-evagination[a]	Maximal evagination
α-Ecdysone	2.15×10^{-5} M	4.3×10^{-5} M
β-Ecdysone	1.0×10^{-7} M	2.1×10^{-7} M
Cyasterone	9.6×10^{-8} M	1.9×10^{-7} M
Inokosterone	1.0×10^{-6} M	2.1×10^{-6} M
Rubrosterone	3.8×10^{-3} M	Not achieved

[a] Degree of evagination is 50% of maximal.
Data from CHIHARA et al. (1972) based on mass isolated disks cultured in ROBB's medium.

the *in vitro* responses to different ecdysones be within an order of magnitude of 3×10^{-7} M. The concentrations of different ecdysones found to induce evagination of *Drosophila* leg disks in ROBB's culture medium (CHIHARA et al., 1972) are presented in Table 3 and the structures of the different molecules in Fig. 5. As can be seen from

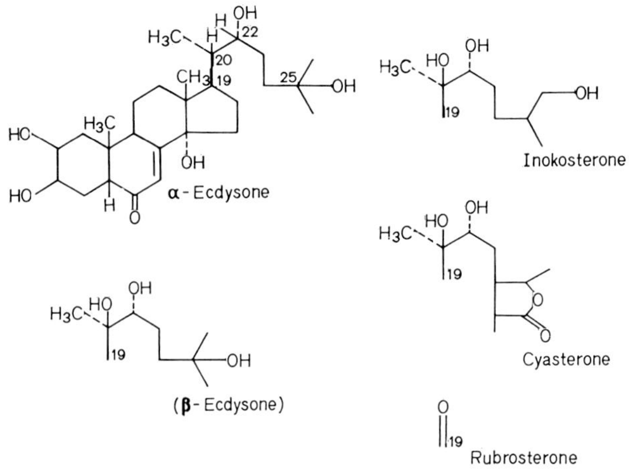

Fig. 5. Structures of α- and β-ecdysone and three phytoecdysones used in the experiments described in Table 3

the table the most active forms all contain an hydroxyl group on carbon 20. Of the two ecdysones found in insects β-ecdysone is by far more active than α-ecdysone. A concentration of 4.3×10^{-5} M of α-ecdysone is required to induce maximal evagination, while a concentration of β-ecdysone of only 2.1×10^{-7} M is needed. MANDARON (1971) reports that in the medium which he has developed, α-ecdysone induces maximum evagination at a concentration of 7×10^{-6} M (about 6 fold less than we have found to be required). However, he achieves complete evagination of *Drosophila* disks, at this concentration of hormone, only when the disks are recovered from larvae about to form puparia. Disks recovered from progressively younger larvae showed progressively less response to α-ecdysone. Our studies, referred to above, were conducted in ROBB's medium using mass isolated disks which, in general, would have been recovered from younger larvae then the oldest larvae used in MANDARON's experiment. MANDARON found that disks from larvae $6^1/_2$ to $6^3/_4$ days old[2] give only partial evagination (about $1/_3$ of maximal) in response to a concentration of α-ecdysone of 7×10^{-6} M. At similar concentrations of α-ecdysone, using mass isolated disks, we also find partial evagination (CHIHARA et al., 1972). This suggests that these disks may be at a comparable developmental stage to MANDARON's $6^1/_2$–$6^3/_4$ day old disks. However, the differences in response found by MANDARON (1971) and in our laboratory may be due to other factors involving differences in the *Drosophila* stocks or culture media used. It is clear, however, that β-ecdysone is substantially more active than α-ecdysone. Similar differences in the activity of these two hormones have also been found by ASHBURNER (1970) for the *in vitro* induction of puffs in polytene chromosomes of *Drosophila* salivary glands. Thus, a concentration of β-ecdysone of approximately 10^{-6} M produces the same sized puff that is achieved by a concentration of α-ecdysone of 10^{-4} M.

Based on the facts that 1. the concentration of β-ecdysone required to induce *in vitro* evagination of *Drosophila* disks (2.1×10^{-7} M) is virtually identical to ecdysone titres found in other insects at molting; 2. β-ecdysone is far more active *in vitro* than α-ecdysone in the production of puffs and evagination of disks and 3. the metabolism *in situ* in *Bombyx* favors the accumulation of β-ecdysone in mature last instar larvae it is very likely that molting hormone activity *in situ* is attributable to β-ecdysone in *Drosophila* and perhaps in many other insects. Supporting this possibility is the observation by POSTLETHWAIT and SCHNEIDERMAN (1969) that disks cultured *in vivo* in adult abdomens undergo complete differentiation in response to β-ecdysone. Since the conversion of β-ecdysone to α-ecdysone has not been detected (KING, personal communication) it is likely that the β-form of the hormone causes complete differentiation of *Drosophila* disks.

Also of interest in the studies by POSTLETHWAIT and SCHNEIDERMAN (1969) is the fact that immense quantities of β-ecdysone were required to achieve evagination *in vivo*. Thus, a concentration approximately 2.7×10^{-4} M was required to induce evagination (about 1000 fold higher than *in vitro*). As these authors pointed out, the immense quantity needed is probably a result of the metabolism of the hormone *in vivo*. Possible metabolism of ecdysones by disks *in vitro* has been investigated using isotopically labeled α- and β-ecdysones. Neither of these compounds are metabolized by the disks in detectable amounts over a 2 h period (CHIHARA et al., 1972). However, ^{3}H-β-ecdysone is rapidly taken up by the disks. SIEGEL (personal communication) has found that maximal binding to disks, approximately 13% of the added hormone, occurs within 1 min after the addition of ^{3}H-β-ecdysone to ROBB's culture medium

---
[2] Under the conditions of axenic growth used by MANDARON, puparium formation occurs about 7 days after egg deposition.

(20000 disks/ml, ^{3}H-β-ecdysone concentration of 6.7×10^{-8} M, 25° C). However, no studies have yet been conducted on the distribution of the hormone in disks. Since ^{3}H-α-ecdysone has been reported to enter larval epidermal cells of *Calliphora* (KARLSON et al., 1964) we presume ecdysone also enters disk cells.

B. Juvenile Hormones

The juvenile hormones isolated from *Hyalophora cecropia* (RÖLLER et al., 1967; MEYER et al., 1970) and some juvenile hormone analogs block the ecdysone-induced evagination of imaginal disks (FRISTROM et al., 1969; CHIHARA et al., 1972). The amount of juvenile hormone required to elicit a supernumerary molt in different insects ranges from 0.005 µg/gm to 90 µg/gm of insect, but generally about 1—2 µg/gm of insect is required (MEYER et al., 1970). However, as can be seen in Table 4, the

Table 4. Inhibition of ecdysone-induced evagination with juvenile hormones

Substance	Half-inhibition (µg/ml)	Complete inhibition (µg/ml)
Cecropia juvenile hormones	75	90
Law-Williams compound	900	1000
Farnesol	1	5

Data from CHIHARA et al. (1972) based on mass isolated disks cultured in ROBB's medium. Ethyl oleate, methyl stearate and farnesenic acid were inactive at concentrations up to 5 mg/ml.

concentrations of Cecropia juvenile hormones required to inhibit evagination of disks are very high (about 90 µg/ml) and even a higher concentration of the crude synthetic juvenile hormone mimetic (LAW-WILLIAMS compound) described by LAW and his colleagues (LAW et al., 1966) is required. The most active analog tested is farnesol. Several comments can be made about these data. Juvenile hormones are oils and are only partially miscible with the culture medium. Thus, the amount of hormone available to the disks is certainly far less than the amount added to the culture medium. The most active compound, farnesol, is the only alcohol in the group (the others being esters) and is probably more soluble than the other compounds which presumably explains its comparatively high activity. The relative inactivity of the juvenile hormones suggests that *in situ* carrier proteins may be involved in trasnport of juvenile hormones. We have also studied the metabolism of Cecropia juvenile hormone in disks. Surprisingly, it was discovered that disks metabolically degrade juvenile hormones producing juvenile hormone diol as well as other degradative products (CHIHARA et al., 1972). Presumably the metabolic degradation also contributes to the low *in vitro* juvenile hormone activity. It is difficult to assess the significance of the ability of disks to catabolize juvenile hormones. It is interesting that mature larval disks, destined to develop into adult structures, have the capacity to destroy the molecules which inhibit that development. Furthermore, it will be interesting to discover whether this capacity exists in disks of all ages, or is acquired just prior to metamorphosis.

V. Macromolecular Synthesis in Disks

A. General Comments

The fact that disks are able to undergo normal development *in vitro* (MANDARON, 1971) indicates that the study of macromolecular synthesis under *in vitro* conditions can produce biologically meaningful results. However, most studies which have been conducted to date on *Drosophila* disks utilize mass isolated disks and conditions under which complete differentiation has not been shown to occur. Some general comments should also be made about the interpretation of data on incorporation of precursors into macromolecules in tissues. The net rate of incorporation of a precursor into a macromolecule, while obviously requiring synthesis of the molecule, is affected by a variety of other factors including the rate of uptake of precursor into the cell, its metabolism and distribution in the tissue and the stability of the macromolecule being formed. Thus, the determination of the rate of synthesis (as opposed to the rate of incorporation) by living cells represents an immensely difficult task. Therefore, increased incorporation into a macromolecule, although frequently interpreted as resulting from increased synthesis, may result from a variety of causes.

B. DNA Synthesis

There is evidence that DNA synthesis in disks occurs throughout the late larval instar in some insects. A paper by KRISHNAKUMARAN et al. (1967) indicates that the synthesis of DNA (as judged by autoradiography, following *in vivo* incorporation of ^{3}H-thymidine) occurs throughout the 4th instar of *Samia cynthia ricini* and *Antheraea polyphemus*. These authors also noted that the amount of incorporation did not change during molts and therefore was not correlated with changes in ecdysone titre. Cell lineage studies on disks in *Drosophila* (BECKER, 1957; BRYANT, 1970; BRYANT and SCHNEIDERMAN, 1969; GARCIA-BELLIDO and MERRIAM, 1971) demonstrate that cell divisions occur throughout the 3rd instar. In some instances cell division stops or is reduced at puparium formation (eye disks, BODENSTEIN, 1940; possibly leg disks, BRYANT, 1970) although it continues for about 18 h after puparium formation in wing disks (GARCIA-BELLIDO and MERRIAM, 1971). Since, in some disks, cell division becomes limited after puparium formation, and the stimulus for puparium formation is ecdysone, then the possibility exists that ecdysone may inhibit DNA synthesis in *Drosophila* disks (in contrast to the situation in moths noted above). This possibility is further supported by the observations of GARCIA-BELLIDO and MERRIAM (1971) and BECKER (1957) indicating that cell divisions in wing and eye become reduced at the time of larval molts.

Evidence from *in vitro* experiments is conflicting. The incorporation of prescursors into disk DNA occurs during initial phases of disk culture but then ceases in both *Drosophila* (ROBB, 1969) and *Galleria* (OBERLANDER, 1969a) wing disks. OBERLANDER (1969a, b) has found that α-ecdysone reinitiates incorporation into DNA. However, he also finds that β-ecdysone does not stimulate incorporation into DNA. A more detailed analysis of the differential effects of α- and β-ecdysone upon incorporation into DNA in *Galleria* disks is presented by OBERLANDER in this volume. The suggestion that α-ecdysone stimulates DNA synthesis in wing disks of *Galleria* seems somewhat at odds with the early observations by OBERLANDER and his co-workers (KRISHNAKUMARAN et al., 1967) in which they suggested that incorporation into DNA in disks of moths was independent of ecdysone titres.

In *Drosophila*, we have investigated the effects of α- and β-ecdysone on incorporation of ^{3}H-thymidine into DNA during the early phases of disk culture (CHIHARA et al., 1972). Both α- and β-ecdysone stimulate incorporation into DNA at concentrations of the respective molecules which induce disk evagination. Thus, during the early phases of disk culture (when incorporation still occurs in the absence of hormone) we find no differential activity between the two ecdysones. We have no information as to the basis of ^{3}H-thymidine incorporation in disks (i.e. whether it represents S-phase synthesis, DNA repair, or some other synthetic process) or to the specific cause of the stimulation in incorporation produced by ecdysone. The experiments performed upon *Drosophila* and *Galleria* disks involve different protocols and distantly related organisms. There is no *a priori* reason to suppose the different results are in conflict. The results from OBERLANDER's laboratory suggesting different functions for α- and β-ecdysone (both stimulating morphogenesis, but only α-ecdysone affecting DNA synthesis) are provocative and may presage new understandings of the regulatory actions of different ecdysones.

C. RNA Synthesis

1. Kinetics of Precursor Incorporation

The original studies on RNA synthesis in disks indicated that the incorporation of ^{3}H-uridine was linear for up to 10 h in SCHNEIDER's medium (FRISTROM and KNOWLES, 1968) and up to 48 h in ROBB's medium (ROBB, 1969). However, the data in these two papers suggest that the rate of incorporation may change rapidly during the first 3—4 h of incubation *in vitro*. RAIKOW (RAIKOW and FRISTROM, 1971), using pulse labeling at different times after the disks were introduced into culture, demonstrated that the rate of incorporation decreases rapidly until reaching a steady state after about 4 h of culture. Since the initial rate of incorporation is restored by β-ecdysone, but never exceeded (see below) it seems possible that mass isolated disks had been exposed *in situ* to ecdysone. However, disks isolated from mid-third instar larvae (when ecdysone titres should be low) showed the same pattern of incorporation as disks isolated from late 3rd instar larvae (when ecdysone titres should be high). Thus, unless larval hormone physiology in *Drosophila* is different than that found in the Lepidoptera, it seems likely that the high level of incorporation during the initial phases of culture is not attributable to *in situ* exposure to ecdysone.

2. Base Composition of Newly Synthesized RNA

The base composition of disk RNA synthesized *in vitro* was determined using $^{32}PO_4$ (FRISTROM et al., 1968). The base composition of RNAs with sedimentation values greater than 5—6 s is similar to the base composition of *D. melanogaster* DNA and rRNA (both with a GC composition of about 40%). Material which sediments with a velocity of 4 s, after short periods of labeling (1 h) has a GC composition of about 45% and after 5 h a GC composition of about 49%. Since the GC composition of *D. melanogaster* tRNA is about 55% (RITOSSA et al., 1966) material sedimenting in the 4s region appears to be a mixture of tRNA and RNA with a low GC composition.

3. Stability of Newly Synthesized RNA

The stability of *in vitro* synthesized RNA has been studied in two ways. Incubation of disks with a 1000 fold excess of ^{1}H-uridine after a 1 h incubation with ^{3}H-uridine did not reduce the amount of incorporated label found in the RNA (FRISTROM et al., 1968). However, blockage of further RNA synthesis with Actinomycin D (1 µg/ml) following different periods of labeling with ^{3}H-uridine, results in the loss of label from the RNA (FRISTROM et al., 1968; RAIKOW and FRISTROM, 1969; STEWART, 1971). The rate at which the ^{3}H-RNA breaks down is in part a function of the duration of incubation with ^{3}H-uridine before the addition of Actinomycin D. Thus, the rate of breakdown is more rapid after a short pulse with ^{3}H-uridine (30 min) than after a long pulse (90 min) (Fig. 6). The data presented in Fig. 6 do not take into account residual incorporation into RNA in the presence of Actinomycin D (only about 95% inhibition is achieved). When corrections for this incorporation are made, the half-life of

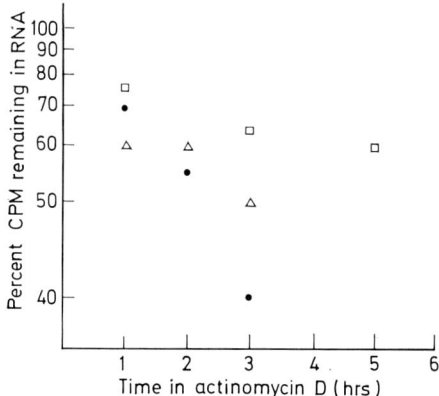

Fig. 6. Stability of newly synthesized RNA. Disks were incubated in Robb's medium at 25° C with ^{3}H-uridine for 30 min (●); 60 min (△) and 90 min (□) and then in the presence of Actinomycin D (1 µg/ml) for varying periods of time. ^{3}H-RNA was determined on the basis of acid-insoluble, base-hydrolysable counts (Data from FRISTROM et al., 1968; RAIKOW and FRISTROM, 1971; and STEWART, 1971)

the RNA synthesized during a 1 h incubation with ^{3}H-uridine, is about 1.5 h (FRISTROM et al., 1968). Since about 85% of the newly synthesized RNA is found in the nuclei after 1 h of labeling (FRISTROM et al., 1968) it is clear that the RNA is breaking down in the nucleus. This phenomenon is characteristic of eukaryotic cells. The fact that breakdown is not detected in the presence of excess ^{1}H-uridine presumably results from preferential reincorporation of ^{3}H-nucleotides following degradation of the ^{3}H-RNA.

4. rRNA Synthesis

It is possible to demonstrate by a variety of techniques that the 28 and 18s rRNAs synthesized in *D. melanogaster* disks in ROBB's medium are qualitatively indistinguishable from those synthesized *in vivo*. Thus, rRNA molecules synthesized *in vivo* and *in vitro* sediment with equivalent velocities, co-chromatograph on benzoylated naphthoylated DEAE-cellulose and electrophorese through acrylamide gel at indistinguishable rates (PETRI et al., 1971). Similar pre-rRNA molecules are synthesized in culture and *in vivo*. The 28 and 18s rRNAs are synthesized from a 38s precursor (FRISTROM et al., 1969; PETRI et al., 1971) as is typically found in insects (EDSTROM and DANEHOLT, 1967). The half-life of the 38s precursor is about 9 min at 25° C in disks cultured in ROBB's medium. This is similar to the 4—5 min half-life of the 45s precursor found in HeLa cells (DARNELL, 1968). The synthesis of rRNA in disks *in vitro* is, however, abnormal in the absence of ecdysone. There is a deficiency in the production of 18s rRNA and the rate of conversion of the precursor into the mature rRNA molecules appears to be slow.

5. Hybridization Characteristics of Newly Synthesized Disk RNA

Drosophila disk RNA synthesized *in vitro* has been characterized using DNA/RNA hybridization techniques. A study of the dependence of the hybridization reaction upon the concentration of ^{3}H-RNA extracted from disks indicates that no clearly definable plateau is achieved over the range of RNA concentrations used (Fig. 7). This result indicates that the sites available on the DNA for forming hybrids have not been saturated. The kinetics of the hybridization reaction indicate that (although the rate becomes substantially reduced after about 24 h) hybrids are still being formed

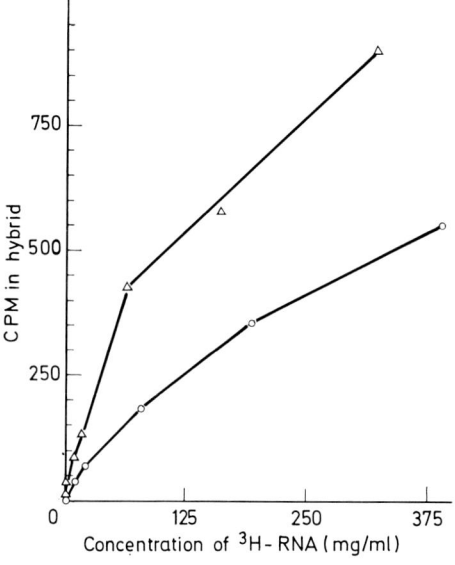

Fig. 7. Concentration dependence of ^{3}H-RNA/DNA hybrid formation with ^{3}H-RNA from disks incubated at 25° C in ROBB's medium with ^{3}H-uridine for 30 min after 2 h incubation with β-ecdysone (△) and without β-ecdysone (○). Disks were fixed in 95% ethanol and RNA extracted using SDS-phenol/cresol. The hybridization was performed in 0.2 ml of 2 × SSC at 61° C for 17 h using filters containing 6.6 μg of homologous DNA/filter. RNA from β-ecdysone treated disks had a specific activity of 1040 CPM/μg, that from control disks: 475 CPM/μg (FRISTROM, unpublished)

even after 40 h of incubation (Fig. 8). The hybrids formed have melting temperatures about 15° C below that of native *D. melanogaster* DNA and are therefore non-specific. Such non-specificity indicates that the hybrids formed involve repetitious sequences of the DNA (BRITTEN and KOHNE, 1968). Therefore, hybrids are formed not with the genes from which the RNA was transcribed, but with DNA sequences which are sufficiently complementary to allow the formation of a hybrid under the conditions

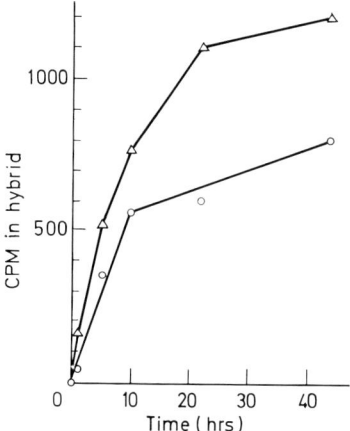

Fig. 8. Kinetics of ^{3}H-RNA/DNA hybrid formation with ^{3}H-RNA from disks incubated at 25° C in ROBB's medium with ^{3}H-uridine for 60 min after a 60 min preincubation with (△) and without β-ecdysone (○). RNA was extracted as in Fig. 7 and the conditions for hybrid formation were similar. RNA (specific activity 1270 CPM/μg) from β-ecdysone treated disks was hybridized at a concentration of 75 μg/ml and the RNA (sp. act. 640 CPM/μg) from control disks at 88 μg/ml. The filters contained 10 μg DNA/filter (FRISTROM, unpublished)

used. The RNAs synthesized in disks are, to some degree, different than those synthesized in other tissues. Experiments depicted in Fig. 9 show the effectiveness of RNAs isolated from different stages during *Drosophila* development as competitors for the formation of hybrids by disk ^{3}H-RNA. As can be seen the most effective competitors are unlabeled disk RNA and adult RNA. Since the sequences of the competitor RNA extracted from disks should be equivalent to the ^{3}H-RNA extracted from disks it is not surprising to find that the disk RNA is the most efficient competitor. The effectiveness of the adult RNA as a competitor is surprising, for although the disks give rise to the adult tissues there would be little expectation to find similar RNAs synthesized in embryonic and adult cells. However, it should be kept in mind that the similarity of the competition curves obtained with disk and adult RNA does not necessarily result from the presence of identical populations of competitor RNAs since the hybrids formed involve repetitious sequences. With the exception of adult RNA, the effectiveness of other competitors follows expectations. Thus, pupal RNA is a better competitor than larval RNA. Furthermore, larval RNA isolated from wild type larvae is a better competitor than that isolated from mutant larvae (X-34) which lack most imaginal tissue (STEWART et al., 1972). Finally, over the concentration range used, embryonic RNA is the least effective homologous RNA tested. However, at higher concentrations than used embryonic RNA may further reduce the amount of hybrid formed between DNA and ³H-RNA isolated from disks. These results indicate that some RNA sequences synthesized in disks are not formed in other *Drosophila* tissues. This is, of course, not surprising.

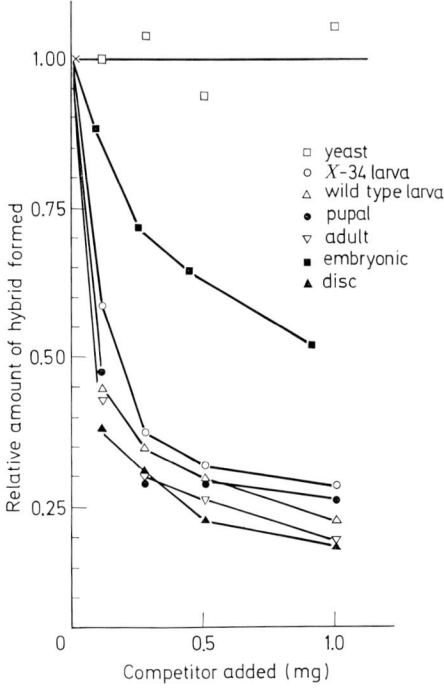

Fig. 9. Effect of different RNAs on the formation of ³H-RNA/DNA hybrids with ³H-RNA from disks incubated for 30 min with ³H-uridine at 25° C in Robb's culture medium. Conditions of hybridization were similar to those described in Fig. 7. Concentration of ³H-RNA (sp. act. 684 CPM/μg) from disks was 105 μg/ml in a volume of 0.2 ml. Homologous filters contained 6.6 μg of DNA/filter. Note that while all of the *Drosophila* RNAs are effective competitors, that yeast RNA does not compete with the ³H-RNA from disks for hybrid formation (Fristrom, unpublished)

D. The Effect of Ecdysone on RNA Synthesis

1. *Rate of Incorporation of Precursors into RNA*

The effect of ecdysone upon the puffing pattern of polytene chromosomes in *Drosophila* salivary glands is well known (Berendes, 1967) and therefore it would not be surprising to discover that ecdysones increase RNA synthesis in disks. An increase in incorporation into RNA following injection of α-ecdysone was documented for epidermal wing tissue of Cecropia pupae by Wyatt (1968). Patel and Madhavan (1969) demonstrated that ecdysone injected into Ricini silkworms, particularly when *in situ* ecdysone titres were low, caused increased incorporation into RNA during subsequent *in vitro* incubation of wing disks. Exposure of *Drosophila* disks *in vitro* to β-ecdysone causes increased incorporation of labeled precursors into RNA (FRISTROM et al., 1969; RAIKOW and FRISTROM, 1971). A significant increase in incorporation is found at a concentration of β-ecdysone of 2.1×10^{-7} M and the maximum increase (relative to the control) is achieved at a hormone concentration of 2.1×10^{-6} M. However, since the action of individual genes cannot be visualized in disks, as it can be in salivary glands, it is difficult to determine the cause(s) of the increased incorporation. Three parameters which might affect the rate of incorporation of precursors into RNA were analyzed (RAIKOW and FRISTROM, 1971). These parameters are 1) the uptake of precursors into disk cells, 2) the stability of the newly synthesized RNA and 3) the distribution of label between the different mono-, di- and trinucleotides. Disks which were preincubated in the absence of ecdysone until the stable basal level of incorporation was reached (see above) were used. β-ecdysone causes a 5—6 fold increase in incorporation of precursor into RNA, but only a 3 fold increase in the uptake of precursor into disk cells. Thus, the increase in uptake does not account for the increase in incorporation into RNA. There is a slight increase in the stability of the RNA synthesized in the presence of β-ecdysone which we have attributed to increased synthesis of 18 and 28 s rRNA (see below). β-ecdysone has no demonstrable effect upon the distribution of label between guanosine mono-, di-, and trinucleotides and therefore the increased incorporation cannot be attributed to increased availability of labeled trinucleotides. Since the increased incorporation of labeled precursors into RNA cannot be totally accounted for by the above considerations, it is likely that β-ecdysone increases RNA synthesis in disks.

The response to β-ecdysone is rapid. A significant increase in incorporation into RNA is found 30 min after the hormone is added, and the maximal response relative to the control is achieved after 4 h. The increased rate of precursor incorporation is dependent upon the continued presence of β-ecdysone. Removal of the hormone results in a rapid decrease in the rate of incorporation which is kinetically similar to the decrease which occurs when disks are first introduced into *in vitro* culture (see above). The rapid reversibility of the effect of β-ecdysone upon precursor incorporation into RNA indicates that the binding of the hormone to receptor sites is also rapidly reversible.

2. *rRNA Synthesis*

β-ecdysone has dramatic effects upon the incorporation of precursors into rRNA in disks (Fig. 10). There are three demonstrable effects: 1) There is a relative increase in the net incorporation into rRNA molecules; 2) There is an increase in the rate of processing of the 38 s pre-rRNA into 28 and 18 s rRNA; 3) The amount of 18 s rRNA synthesized is dramatically increased and the ratio of 28 s and 18 s rRNA synthesized is the same as that found in mature ribosomes (FRISTROM et al., 1969; PETRI et al., 1971). The synthesis of rRNA in disks cultured *in vitro* is comparable to the *in vivo* synthesis of rRNA in disks (STEWART, 1971) (Fig. 11). Indeed there is the suggestion, although the labeling conditions are obviously different, that processing occurs more rapidly *in vitro* in the presence of hormone, than *in vivo*. The effect of β-ecdysone upon rRNA synthesis is rapid. An increase in incorporation into 18 and 28 s rRNA can be detected within 30 min after the addition of the hormone (Fig. 12) and by extrapolation β-ecdysone presumably affects rRNA synthesis in disks within 20 min after its addition to the culture medium.

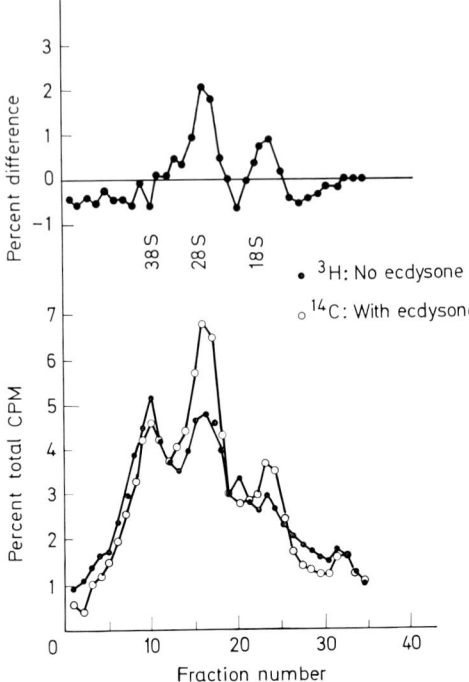

Fig. 10. Effect of β-ecdysone on rRNA synthesis in disks. Disks were incubated for 2 h with (O) and without (•) β-ecdysone and then respectively for 1 h with ^{14}C-uridine and ^{3}H-uridine. RNA was extracted with SDS-phenol/cresol and then sedimented through a 5–20% sucrose gradient. The distribution of counts along the gradient and the difference in percent ^{14}C and percent ^{3}H are depicted. Note that the increases caused by β-ecdysone are restricted to rRNA (Data from PETRI et al., 1971)

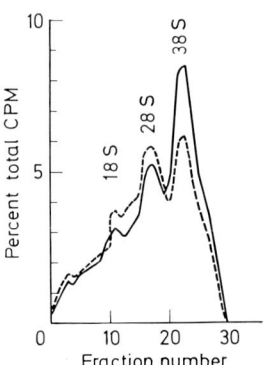

Fig. 11. Comparison of rRNA synthesis in imaginal disks *in vivo* (———) and *in vitro* in the presence of β-ecdysone (- - - - -). Mid third instar larvae were injected with ^{3}H-uridine, kept at 25° C and pulled open in 95% ethanol after 30 min and the disks were subsequently recovered. Disks were pre-incubated with β-ecdysone for 2 h in ROBB's medium at 25° C and were then incubated for 30 min with ^{3}H-uridine. The RNA was extracted with SDS-phenol/cresol and then sedimented through a 5–20% sucrose gradient. Note that the processing of the 38s precursor seems faster *in vitro* with β-ecdysone than *in vivo* (Data from STEWART, 1971)

It is difficult to determine by what mechanisms the increased incorporation into rRNA is achieved. One possibility is that there is increased production of the 38s pre-rRNA which by mass action increases processing of intermediates into mature 18 and 28s molecules. An alternative possibility is that the hormone mainly affects processing of pre-rRNA. In the absence of β-ecdysone not only is the processing of the 38s molecule slow, but also the synthesis of the 18s rRNA is strikingly reduced.

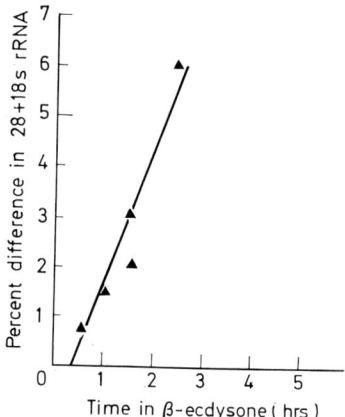

Fig. 12. Increase in net synthesis of rRNA in disks exposed to β-ecdysone. Disks were pre-incubated with β-ecdysone for varying periods of time and then with ^{3}H-uridine for 1 h. The points are plotted at the mid-point of the incubation period with ^{3}H-uridine. For the 30 min point β-ecdysone and ^{3}H-uridine were added at the same time. Percent difference in 28s and 18s rRNA is determined using difference plots similar to that shown in Fig. 10. The difference in percent CPM in the 28s and 18s material synthesized in the presence and absence of β-ecdysone is totalled to determine the percent difference in rRNA synthesis (FRISTROM, unpublished)

One can only conclude that the absence of production of 18s molecules implies that there is breakdown of macromolecular pre-rRNA to nucleotides. Thus, increased incorporation into 18s molecules in the presence of β-ecdysone could result entirely from the increased success in converting the 38s precursor into the mature rRNA molecule and not into nucleotides. Such a condition would result in an increase in the overall synthesis of rRNA, increase the rate of processing of the 38s precursor and increase the amount of 18s rRNA formed. This possibility also suggests that an ecdysone binding protein may be involved in processing of pre-rRNA into mature rRNA.

3. Hybridization of Disk RNA

Hybridization of RNA synthesized in *Drosophila* disks in the presence and absence of β-ecdysone has been studied. The amount of hybrid formed is affected by the hormone. After short incubations with β-ecdysone the hybridizing capacity[3] of

---

3 Hybridizing capacity is defined as follows:

$$\frac{^3\text{H-RNA(CPM) in RNA/DNA hybrid}}{\text{Total }^3\text{H-RNA(CPM) in reaction.}}$$

³H-RNA is increased compared to controls (Fig. 13). The increase in hybridizing capacity is mainly demonstrable at low inputs of ³H-RNA and therefore involves relatively repetitious sequences of DNA (keeping in mind that all hybrids formed involve repeated sequences). The increase in hybridizing capacity is detectable

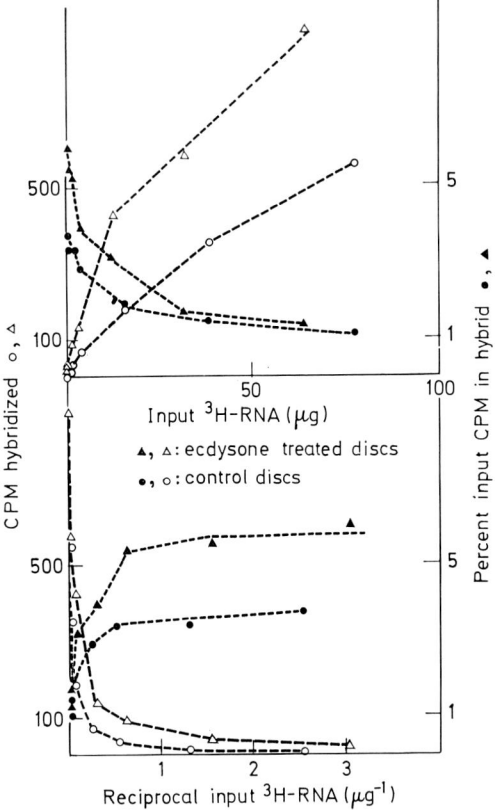

Fig. 13. Effect of β-ecdysone on the hybridizing capacity of disk ³H-RNA. Disks were incubated in ROBB's culture medium at 25° C for 2 h in the presence (△, ▲) or absence of β-ecdysone (○, ●) and then for 1 h with ³H-uridine. RNA was extracted using SDS-phenol/cresol. Hybridization was carried out using the specific conditions described in Fig. 7. Concentration dependence of hybrid formation and the percent input CPM in hybrid (hybridizing efficiency) is depicted. Reciprocal input of ³H-RNA is depicted to demonstrate that at low inputs of RNA the percent which hybridizes is constant. Note that β-ecdysone increases the percentage of input ³H-RNA which forms hybrids at low inputs of RNA, but that at high inputs of RNA the hybridizing capacities are similar (FRISTROM, unpublished)

within 1 h after the addition of β-ecdysone to the culture medium (Fig. 14). This increase, however, appears to be transitory. Preliminary experiments indicate that the hybridizing capacity of ³H-RNA extracted from disks exposed to β-ecdysone for 5 h is lower than that of ³H-RNA extracted from control disks. The increase in hybridizing capacity after a 1 h exposure of disks to β-ecdysone presumably results from an increase of synthesis of RNA molecules which can form hybrids with the DNA under the conditions used. These data, therefore, support the proposal based on incorporation experiments, that β-ecdysone stimulates RNA synthesis in disks.

Competition experiments with RNA from disks incubated in the presence and absence of β-ecdysone were also conducted (Fig. 15). As can be seen in the figure

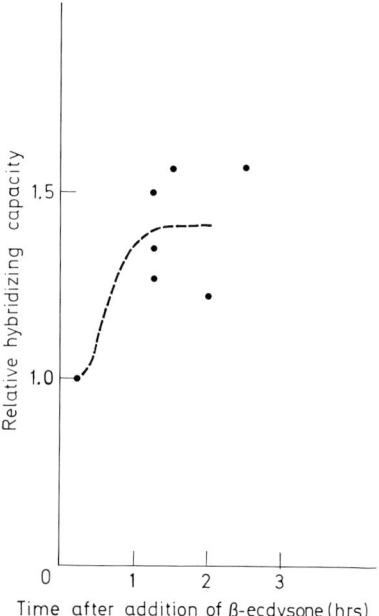

Fig. 14. Time dependence of effect of β-ecdysone on hybridizing capacity of disk ³H-RNA. Disks were incubated in ROBB's culture medium at 25° C with and without β-ecdysone for varying periods of time. RNAs were extracted and hybridized as is described in Fig. 13. Using the reciprocal input plots, comparisons were made between the hybridizing capacity at low inputs of RNAs extracted from disks with and without exposure to β-ecdysone. Relative hybridizing capacity is defined as

$$\frac{\text{Hybridizing capacity of RNA from ecdysone treated disks}}{\text{Hybridizing capacity of RNA from control disks}}$$

the ¹H-RNAs extracted from the two sources decrease the amount of hybrid formed in similar manners. Only one difference in the effectiveness of the competitors has been detected. ¹H-RNA from disks exposed to ecdysone is a more effective competitor at low inputs of ¹H-RNA against the formation of hybrids with ³H-RNA from ecdysone-exposed disks than control RNA. This result, suggesting a quantitative increase in certain types of RNA, is consistent with the data presented above on hybridizing capacity. The absence of demonstrable differences in competition at high inputs of RNA should not be taken as necessarily indicating that there are no differences in the sequences of the RNAs synthesized in the presence and absence of β-ecdysone. First, the data deal only with RNAs which bind to repeated DNA sequences and do not involve RNAs binding to unique sequences. Second, hybrids are formed with DNA sequences which were not used for the transcription of the RNA. Therefore, the competitor ¹H-RNAs can have different sequences than the ³H-RNAs forming hybrids and still prevent the formation of a DNA/RNA hybrid.

A few general comments can be made about the effects of β-ecdysone upon RNA synthesis in disks. All of the effects noted: increased incorporation, increased hybridizing capacity, increased synthesis of 28 and 18s rRNA occur rapidly after the addition

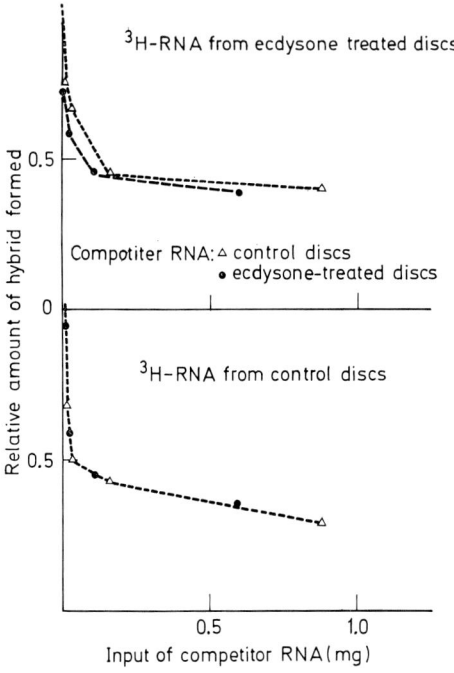

Fig. 15. Effect of RNAs from disks incubated with and without β-ecdysone on the capacity of ³H-disk RNA to form hybrids. Disks were incubated in ROBB's culture medium at 25° C for 2 h with and without β-ecdysone and then with ³H-uridine for 1 h. RNA was extracted with SDS-phenol/cresol. RNA was also prepared from disks incubated with (●) and without (△) β-ecdysone for 3 h, and was used as the competitor RNA in the experiment. The conditions of hybridization were similar to those described in Fig. 7. ³H-RNA from ecdysone treated disks (sp. act. 15480 CPM/μg RNA) was incubated at a concentration of 4.32 μg/ml and ³H-RNA from control disks (sp. act. 7000 CPM/μg RNA) was incubated at a concentration of 4.75 μg/ml. The homologous filters contained 6.6 μg of DNA

of hormone. In all cases the effects are detected within an hour, and in all likelihood β-ecdysone acts even more rapidly. Thus, it is possible that the primary action of β-ecdysone upon disk cells is on RNA synthesis. The fact that β-ecdysone appears to affect RNA synthesis in both the nuclei and in the nucleoli with equal rapidity suggests multiple sites of action for the hormone.

E. The Effect of Juvenile Hormone Upon RNA Synthesis

The fact that the development of imaginal disks is under the control of juvenile hormone has been known for several years (SEHNAL, 1968). Experiments by PATEL and MADHAVAN (1969) using *Samia cynthia* indicate that injection of juvenile hormone mimetics into larvae produces an increase in incorporation into RNA in disks dissected 12 h after the injection and incubated for 4 h *in vitro* with labeled precursor. The juvenile hormone mimetics were most effective when injected during periods when the *in situ* titre of juvenile hormone was low. These authors also reported that the simultaneous injection of ecdysone and a juvenile hormone mimetic produced no increase in incorporation into RNA. Thus, the two hormones were antagonistic to each other.

Studies on the effects of Cecropia juvenile hormones and the synthetic juvenile hormone analog (LAW et al., 1966) on the incorporation of ^{3}H-uridine into RNA in *Drosophila* disks cultured *in vitro* have been conducted in our laboratory. The synthetic juvenile hormone analog has a surprising effect *in vitro* in that it prevents the transport of precursors into disk cells and thus completely blocks incorporation (FRISTROM et al., 1969). The two Cecropia juvenile hormones (RÖLLER et al., 1967; MEYER et al., 1970) do not substantially alter the rate of incorporation of ^{3}H-uridine into disk RNA (CHIHARA et al., 1972). The Cecropia hormones reduce the β-ecdysone-stimulated increase in incorporation into disk RNA and are therefore antagonistic to β-ecdysone. However, reliable studies on the effect of juvenile hormones upon uptake of prescursors into disks have not yet been conducted. Thus, the mechanism(s) by which the juvenile hormones inhibit the β-ecdysone-stimulated increase in incorporation into RNA is not known. It should be kept in mind that excessive quantities of juvenile hormone are needed to inhibit evagination (see above) and therefore the *in vitro* action of the juvenile hormone upon incorporation into RNA may not reflect the normal biological action of the hormone.

F. Protein Synthesis

Only a few studies have been conducted on protein synthesis in disks and most of them are limited to *in vitro* conditions. SDS-acryamide gel electrophoresis of the proteins recovered from the cytoplasm, the ribosomes and a $17000 \times$ g pellet (containing nuclei and cell debris) reveals many different bands and distinctly different distributions of proteins in the three fractions (Fig. 16). A comparison is also made in Fig. 16 between the proteins synthesized *in vitro* and those found in the disks at the time of isolation. It is difficult to make exact comparisons between the continuous optical density trace produced by scanning the stained gels and the distribution of label produced by counting 1 mm segments of the gel. However, it is clear that the distribution of label generally follows the distribution of proteins along the gel. This indicates that the bulk of the protein found in disks is synthesized in disk cells and not sequestered from the hemolymph.

The same conclusion was reached in the case of a known enzyme, alcohol dehydrogenase (ADH) in a series of transplantation experiments carried out by URSPRUNG et al. (1970). It was known that male genital disks contain no ADH activity, while their adult derivatives do. Disks of a particular electrophoretic ADH variant were transplanted into host larvae of a different electrophoretic variant or into ADH-negative larvae. When the implants were recovered from their host flies after metamorphosis and electrophoresed, they were seen to contain ADH corresponding to their own genotype, indicating the enzyme was not accumulated from the host's hemolymph. It was probably synthesized by the disk cells themselves during metamorphosis, although activation cannot be excluded in this experiment.

The incorporation of ^{3}H-serine into protein in disks cultured in Robb's medium is essentially linear for up to 44 h (Robb, 1969) although there are changes in the rate

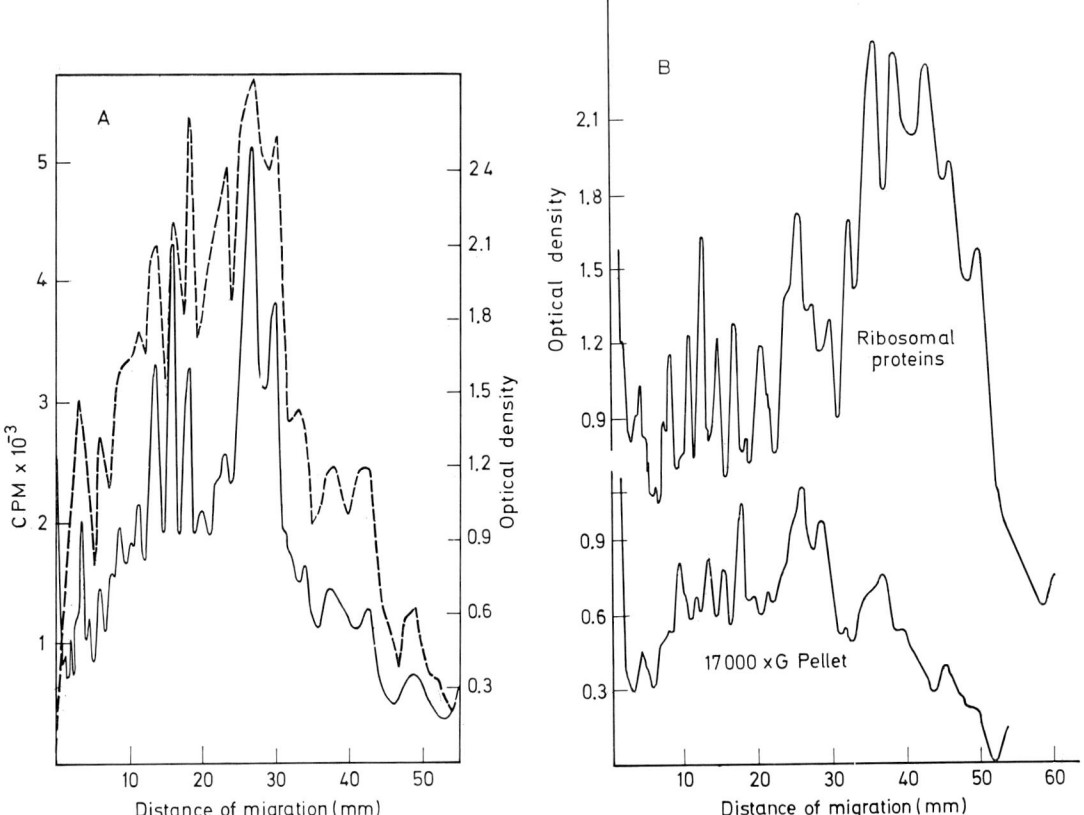

Fig. 16 A and B. SDS-acrylamide gel electrophoresis (7.5% Acrylamide) of proteins isolated from a 17000 × g pellet, ribosomes and the cytosol. Proteins synthesized during a 2 h incubation with ^{3}H-leucine (- - - - - -) are also depicted. Optical density was determined by scanning the stained gel. ^{3}H-protein was determined by slicing the gel into millimeter slices and placing the slices in 10 ml of scintillation fluid containing 0.2 ml of 1% SDS, 1 ml of Nuclear Chicago Solubilizer and 8.8 ml of toluene scintillation cocktail. Proteins in the 17000 × g pellet and in the ribosomes were solubilized by extraction with 8 M urea (Siegel, Gregg, and Fristrom, unpublished)

of incorporation during the early phases of culture. Label incorporated into protein during a 1 h incubation is stable to a "chase" with unlabeled amino acids for at least 5 h (Fristrom et al., 1968). Similar results are obtained when further protein synthesis is blocked with puromycin (unpublished observation). These data indicate that the bulk of the protein synthesized in disks is comparatively stable, but give no information about specific proteins. Based on studies using Actinomycin D (Fristrom et al., 1968) the mRNA of disks has an average half-life of about 2.5 h. This is longer than newly synthesized nuclear RNA which has a half-life of 1—1.5 h (see above). There are, however, indications of variation in the half-lives of disk mRNAs. First, 5—10% of the incorporation of labeled leucine is stable to Actinomycin D poisoning over prolonged periods of time. Thus, some mRNAs are stable and continue to

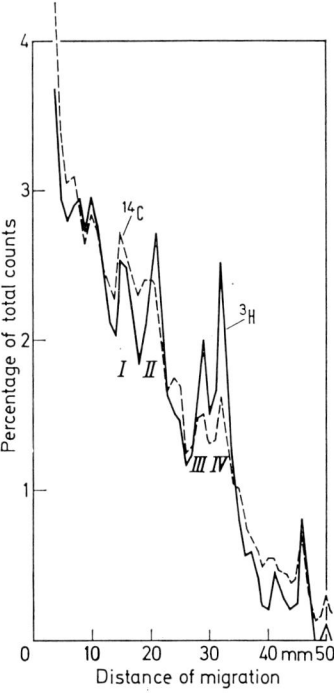

Fig. 17. Electrophoretic pattern of labeled proteins isolated from disks treated and untreated with Actinomycin D (1 μg/ml) for 15 h were incubated with ^{3}H-leucine while disks incubated only in tissue culture medium for 15 h were incubated with ^{14}C-leucine. Roman numerals identify positions of major protein bands in the gel (FRISTROM et al., 1968)

function under conditions in which there is no further synthesis of high molecular weight RNA. Second, gel electrophoresis of proteins synthesized before and after Actinomycin D poisoning demonstrates that some proteins are synthesized in comparatively greater amounts after Actinomycin D poisoning (Fig. 17). This result also indicates that some RNAs are more stable than others since if all mRNAs were equally stable there should be no change in the relative labeling of the different protein fractions. The Actinomycin D-resistant synthesis in disks apparently involves protein species present in relatively large concentrations (FRISTROM et al., 1968), and indicates that these proteins are synthesized on stable mRNAs. GROSS and COUSINEAU (1963) had previously found, during early cleavage of sea urchin eggs, that the synthesis of microtubular protein involved the use of comparatively stable mRNA. Similarly KAFATOS (1969) has demonstrated that the synthesis of large quantities of cocoonase in the galeae of silk moths involves the use of a highly stable mRNA with a half-life greater than 40 h. The bulk of mRNAs in the galeae have half-lives of about 1—2 h.

G. Effect of Ecdysone on Protein Synthesis

Ecdysone increase incorporation of amino acids into proteins in disks and imaginal tissues. WYATT (1968) demonstrated that α-ecdysone causes increased incorporation of leucine into proteins of wing tissue of Cecropia silk moth pupae. PATEL and MADHAVAN (1969) demonstrated that both α- and β-ecdysone injected into the RICINI silkworm caused increased incorporation of ^{14}C-protein hydrolysate into disks dissected 12 h after injection of the hormone and incubated *in vitro* with the precursors for 4 h. These authors pointed out that the greatest stimulatory effect was found when the hormones were injected at a stage when the *in situ* ecdysone titre was low. β-ecdysone induces increased incorporation of ^{3}H-leucine into protein of *Drosophila* disks incubated *in vitro* (Table 5). As can be seen in the table incorporation does

Table 5. Effect of β-ecdysone on protein synthesis in imaginal disks

Period of pre-incubation with β-ecdysone (in h)	Incorporation into protein		Increase significant at 5% level
	With β-ecdysone (CPM/μg RNA)	Without β-ecdysone (CPM/μg RNA)	
0	96.1	103.8	—
1	160.4	151.6	No
2	179.8	155.5	No
3	215.6	164.2	No
4	237.9	152.1	No[a]
5	228.8	131.7	Yes
7	225.2	119.8	Yes[b]
10	208.8	104.3	Yes[b]

Data are the average of three experiments. Labeling period was for 30 min with ^{3}H-leucine following preincubation period. Incorporation was determined on basis of acid-insoluble, base-non-hydrolysable counts expressed in terms of amount of RNA present in the acid precipitate.
[a] Significant at 10% level.
[b] Significant at 1% level.
Data from SIEGEL, GREGG, and FRISTROM (unpublished).

not change dramatically relative to the control during the first 2 h of incubation with β-ecdysone, but has significantly increased 5 h after the addition of the hormone. Thus, β-ecdysone stimulates increased incorporation into protein as well as RNA and DNA. However, there is a greater time lag between the addition of the β-ecdysone and the increase in incorporation into protein than that found for RNA (see above). The length of the lag suggests that β-ecdysone does not act directly upon the protein synthesizing machinery but produces the increased incorporation indirectly (possibly through increased production of mRNA). Also, unlike the effect of ecdysone on incorporation into RNA which is dependent upon continued presence of the hormone, once the increased level of incorporation into protein is achieved, it is maintained after β-ecdysone is removed (Table 6). This observation also indicates that β-ecdysone does not directly affect protein synthesis. Associated with the increased incorporation into protein is an increase in the amount of rough endoplasmic reticulum found in disk cells. Several authors (URSPRUNG and SCHABTACH, 1968; FRISTROM, 1969b;

Table 6. Effect of removing β-ecdysone on protein synthesis in disks

Period of incubation after removing β-ecdysone (in h)	Incorporation into protein	
	With β-ecdysone preincubation (CPM/μg RNA)	Without β-ecdysone preincubation (CPM/μg RNA)
0	280.6	159.4
2	232.4	161.3
4	251.9	149.6
6	263.5	137.8

Data represent the average of 3 experiments. Protocol was similar to that described in Table 5. Disks were preincubated for 6 h prior to the start of the experiment. Data from SIEGEL, GREGG, and FRISTROM (unpublished).

CHIARODO and DENYS, 1968; POODRY and SCHNEIDERMAN, 1970) have noted increases in the amount of rough endoplasmatic reticulum in disk cells following puparium formation (also see the article by URSPRUNG in this volume). We have reported (FRISTROM et al., 1969) that β-ecdysone induces an increase in rough endoplasmic reticulum *in vitro*. However, this study was not exhaustive and should be considered preliminary. Substantial increases in rough endoplasmic reticulum have been found to occur during hormone-induced metamorphosis in bullfrog tadpole liver cells by TATA (1968) who believes that such increases are characteristic of the action of growth and developmental hormones.

We have investigated whether β-ecdysone affects the synthesis of all proteins equally or if increased synthesis of specific proteins was involved. Considering the fact that β-ecdysone causes striking morphogenetic changes in imaginal disks it would not be surprising to discover an increase in the net synthesis of specific proteins. In order to test this possibility disks were incubated with and without β-ecdysone and then labeled with ^{14}C- or ^3H-leucine respectively for 2 h. The two populations of disks were mixed, extracted and three protein fractions consisting of proteins found in the cytoplasm, the ribosomes and the $17000 \times g$ pellet were recovered. The proteins were subjected to SDS-acrylamide gel electrophoresis and the distribution of ^{14}C and ^3H along the gel determined. A comparison of incorporation into disk cytoplasmic proteins with or without a prior exposure to β-ecdysone for 5 h is depicted in Fig. 18. Also shown is a difference plot from an experiment in which neither of the two populations of disks was incubated with β-ecdysone. As can be seen in the figure there is a difference in the distribution of labeled proteins synthesized in the presence of β-ecdysone compared to the control. During the first 3—4 h of incubation in β-ecdysone no dramatic differences are detected in cytoplasmic proteins synthesized in the presence or absence of β-ecdysone. This again demonstrates that the effect of β-ecdysone on protein synthesis is delayed. In the ribosome fraction, increased incorporation into low molecular weight material is found after 2—4 h of exposure to β-ecdysone (Fig. 19).

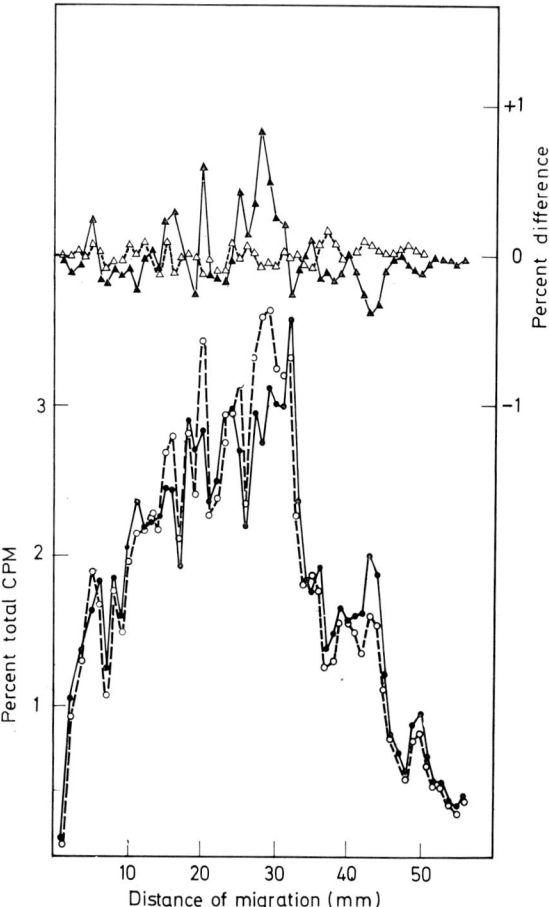

Fig. 18. Effect of β-ecdysone upon protein synthesis. Disks were incubated with (------) or without (———) β-ecdysone for 5 h and then respectively with ^{14}C-leucine and ^{3}H-leucine for 2 h. Proteins were recovered from the cytosol fraction and subjected to SDS-acrylamide gel electrophoresis. The gels were treated as is described in Fig. 16, and the distribution of ^{3}H and ^{14}C along the gel determined. The difference plot (—▲—) shows the difference in the proteins synthesized in the presence and absence of β-ecdysone. For comparison a control is included in which neither of the two populations of disks was exposed to β-ecdysone (—△—) (SIEGEL, GREGG, and FRISTROM, unpublished)

In an earlier study on protein synthesis in disks (FRISTROM and KNOWLES, 1967) in which synthesis in larval and prepupal disks was compared, no major differences were found. The technique used was similar but less refined than the one described above and the failure to detect differences in protein synthesis might have been a technical one. However, the prepupal disks used in the earlier experiment were morphologically distinct from the larval disks. The disks treated with β-ecdysone in the experiments described above, however, are not dramatically different in appearance than those not exposed to β-ecdysone (the first signs of evagination being clearly

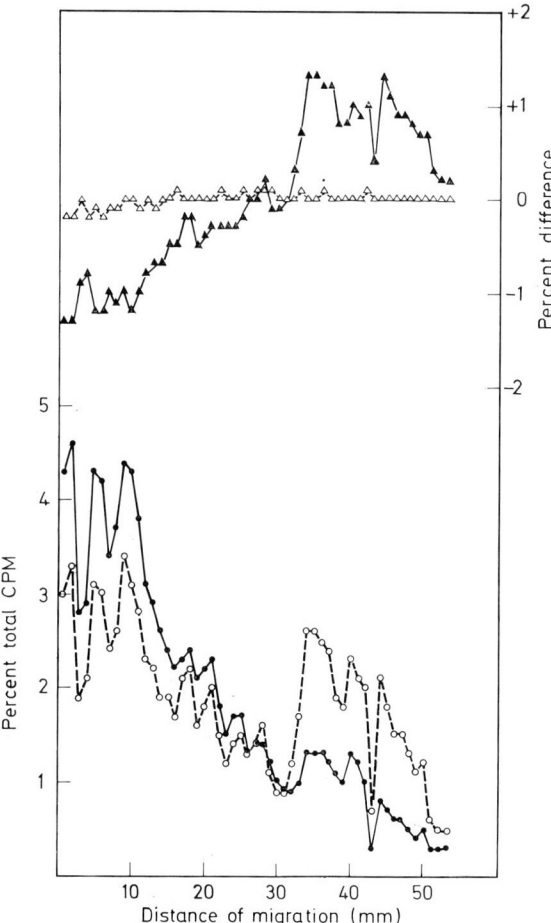

Fig. 19. Effect of β-ecdysone upon synthesis of ribosome associated proteins. Disks were incubated with (-----) and without (———) β-ecdysone for 2 h and then respectively with ^{14}C-leucine and ^{3}H-leucine for 2 h. The ribosomal pellet was first extracted with 1.0 M KCl and was then dissolved in 8 M urea before being subjected to SDS acrylamide gel electrophoresis. The difference plot (— ▲ —) shows the difference in the proteins synthesized in the presence and absence of β-ecdysone. For comparison a control is included in which neither of the two populations of disks was exposed to β-ecdysone (—△—) (SIEGEL, GREGG, and FRISTROM, unpublished)

detected about 6 h after the addition if β-ecdysone). Thus, the possibility also exists that the changes in protein synthesis are transitory and can only be detected during a narrow time period after the addition of β-ecdysone.

H. Effect of Juvenile Hormone Upon Protein Synthesis

Patel and Madhavan (1969) demonstrated that the injection of juvenile hormone analogs into the Ricini silk moth induced increased incorporation of ^{14}C-protein hydrolysate into protein in disks dissected 12 h later and incubated for 4 h in the label. These authors noted that the juvenile hormone analogs had their most pronounced effect when the *in situ* titre of juvenile hormone was low. In addition, it was also noted that ecdysone and the juvenile hormone analogs were antagonistic to each other in their effects when injected simultaneously. Studying *Drosophila* disks *in vitro* Fristrom et al. (1969) demonstrated that the Law-Williams mixture (Law et al., 1966) blocked the uptake of ^{14}C-leucine into disk cells and therefore inhibited its incorporation into protein. Cecropia juvenile hormones (Röller et al., 1967; Meyer et al., 1970) do not noticeably affect the incorporation of leucine into proteins in disks cultured *in vitro* (Chihara et al., 1972). Cecropia hormones do, however, inhibit the β-ecdysone-induced increase in incorporation into protein. Thus, the naturally occurring juvenile hormones from Cecropia inhibit the ecdysone-induced *in vitro* stimulation of incorporation into RNA and protein and also inhibit evagination. These observations clearly demonstrate that juvenile hormone and ecdysone act directly and antagonistically upon synthetic and developmental processes in disks.

VI. Cuticle Formation and Hexosamine Metabolism

Larval imaginal disks have no cuticle. Imaginal tissue in the late prepual period, however, does have a cuticle. The formation of this extracellular cuticle represents one of the major synthetic and secretory activities of disks during the prepual period. Cuticle formation in disks appears to be essentially similar to cuticle formation in other insects (Robson, 1964; Locke, 1966, 1969). The cuticle is synthesized in layers starting with the outermost cuticulin layer which is composed of protein deposited by the microvilli of the disk epidermal cells. Internal to the cuticulin layer is a less dense layer, the epicuticle, which is also composed of protein and clearly present 6 h after puparium formation in *D. melanogaster* (Fristrom, 1969b; Wehman, 1969; Poodry and Schneiderman, 1970). The deposition of the last layer, the endocuticle, is first detected about 7 h after puparium formation (Poodry and Schneiderman, 1970) and by 12 h is clearly visible (Fristrom, 1969b). It is this last layer which contains both protein and chitin (Richards, 1951).

There is good reason to believe that the proteins and the chitin found in the cuticle are synthesized in disk cells and not sequestered from the hemolymph. Mandaron (1971) reports the complete differentiation of *Drosophila* disks *in vitro* in a defined culture medium. Therefore the deposition of the cuticle occurs under conditions in which no extrinsic macromolecules are available. The induction of cuticle formation by β-ecdysone in the wing disks of the rice stem borer *Chilo suppressalis* in culture has also been reported (Agui et al., 1969). The synthesis of cuticle in *Drosophila* disks cultured in Robb's medium has not been detected (Fristrom et al., 1969). The *in vitro* formation of cuticle in disks isolated from members of the Lepidoptera is discussed by Oberlander elsewhere in this volume.

Chitin, which is found only in the endocuticle, is a polymer containing mainly N-acetylglucosamine, but also some glucosamine (Giles et al., 1958). There is good evidence that the polysaccharide is synthesized from uridine diphospho-N-acetylglucosamine (UDPNAGA) (GLASER and BROWN, 1957). *Drosophila* disks from 3rd instar larvae are capable of synthesizing UDPNAGA *in vitro* from at least three sources: glucose, glucosamine and N-acetylglucosamine (FRISTROM, 1968). The synthesis probably occurs via the metabolic pathway depicted in Fig. 20. As can be seen in the figure the conversion of glucose (or any hexose) to UDPNAGA requires

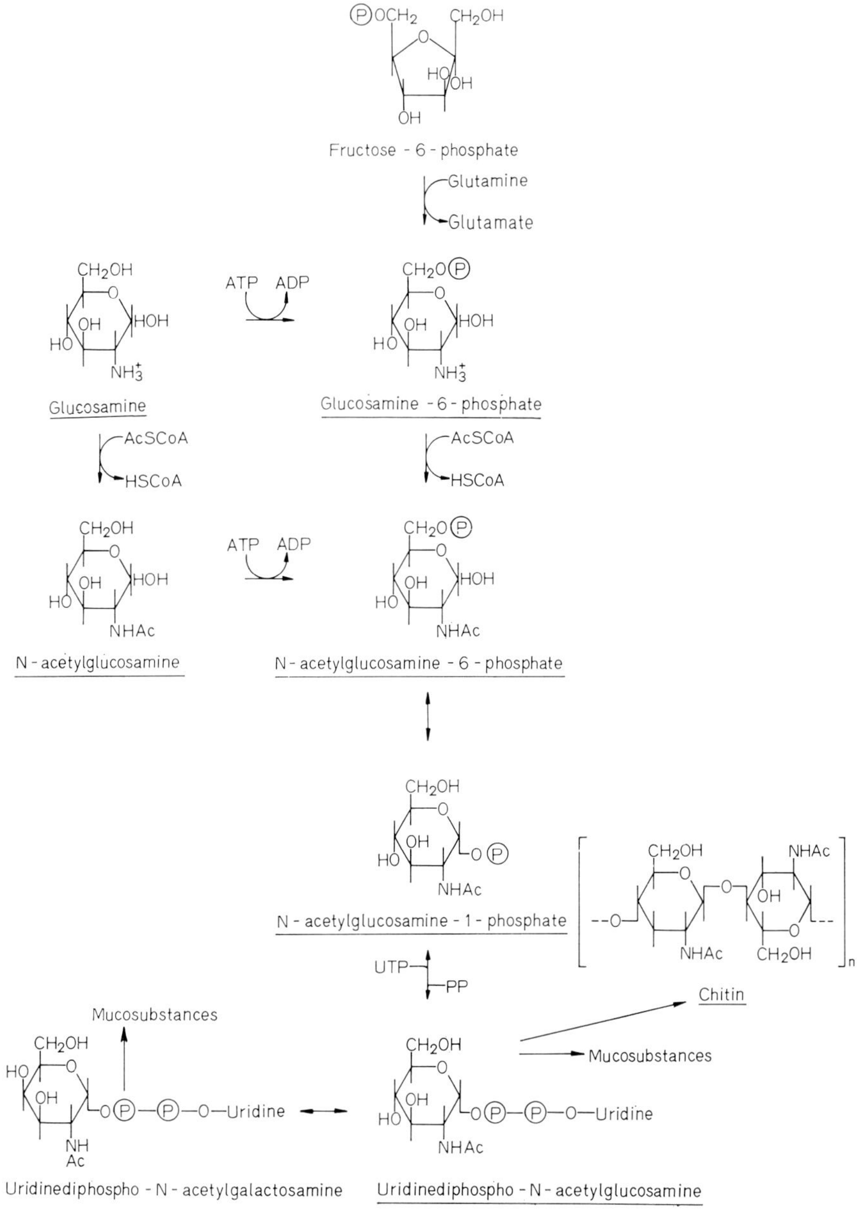

Fig. 20. Major pathways of hexosamine metabolism

an amination involving glutamine to form glucosamine-6-phosphate (GA-6-P). The formation of phosphorylated hexosamines from glucosamine or N-actylglucosamine, however, involves only the phosphorylation of the amino sugars. In terms of the concentration of precursor added to the culture medium, the greatest net synthesis of UDPNAGA is derived from glucosamine and is substantially greater ($\times 50$) than that obtained from glucose. With the exception of the initial metabolic steps a common biosynthetic pathway is shared in the conversion of the two molecules to UDP-NAGA (Fig. 20). Therefore in the formation of UDPNAGA from hexoses (glucose, fructose, trehalose) the rate limiting step appears to be the amination of fructose-6-phosphate to produce GA-6-P. In disks isolated from larvae the rate of net synthesis of UDPNAGA from glucose is about 0.01 mμmoles/1000 disks/h (about 0.01 mμmoles/mg wet weight of tissues/h). In contrast the net synthesis of UDPNAGA from glucosamine is about 0.6 mμmoles/1000 disks/h (about 0.6 mμmoles/mg wet weight of tissue/h). Integuments recovered from pharate pupae at the time of head emergence (12 h after puparium formation at 25° C) contain about 100 mμmoles of glucosamine per mg *dry* weight of tissue (FRISTROM, 1965). Assuming disks are 95% water, then the net rate of synthesis of UDPNAGA from glucosamine is about 12 mμmoles/mg dry weight per hour, and 100 mμmoles of chitin could be produced in about 8 h. However, equivalent synthesis from glucose using the maximal rate detected in larval disks, would require about 500 h. MANDARON (1971) has demonstrated that the synthesis of cuticle and therefore chitin occurs *in vitro* in a medium containing no hexosamines. Therefore, the rate limiting step in the formation of hexosamines from hexoses (the conversion of fructose-6-phosphate to glucosamine-6-phosphate) must be amenable to a regulated increase in activity. However, with the exception of this single rate limiting step it is clear that even larval disks are capable of synthesizing large quantities of UDPNAGA and are therefore prepared to synthesize large quantities of chitin during prepupal development. This observation is a good example of preadaptation of disks for the high synthetic demands which are made during the rapid development of disks during metamorphosis.

VII. The Mechanism of Disk Evagination

An understanding of the molecular basis of evagination is a necessary prerequisite for investigations on the regulatory mechanisms involved. The process of evagination involves an increase in the surface area of the appendage (Fig. 21). The possibility that this increase depends upon cell division is unlikely. Although there are cell divisions in the wing disk during the prepupal period (STUMPF, 1956) these are not concentrated in the wing Anlage but occur throughout the disk (GARCIA-BELLIDO and MERRIAM, 1971). The number of cell divisions in the leg disk appears to be limited after puparium formation (BRYANT and SCHNEIDERMAN, 1969). Furthermore, evagination *in vitro* is not inhibited by colchicine (see below). In a recent paper, MANDARON (1971) emphasizes that evagination is not caused by extrinsic forces (e.g. hydrostatic pressure) but appears to be an intrinsic characteristic of disk cells. This finding is in agreement with our proposal (FRISTROM et al., 1969) that evagination results from a change in shape of disk cells. It has long been known that the majority of cells found in the appendage Anlagen of late larval disks are columnar (AUERBACH, 1936; and recently observed by FRISTROM, 1969b; POODRY and SCHNEIDERMAN, 1970). However, the cells in the appendages which have completed evagination are cuboidal and at least in the wing disk become highly flattened at the end of the prepupal period (FRISTROM, 1969b; POODRY and SCHNEIDERMAN, 1970). This change in cell shape results in a substantial increase in the surface area of the disk tissue (Fig. 21) (FRISTROM, 1969b).

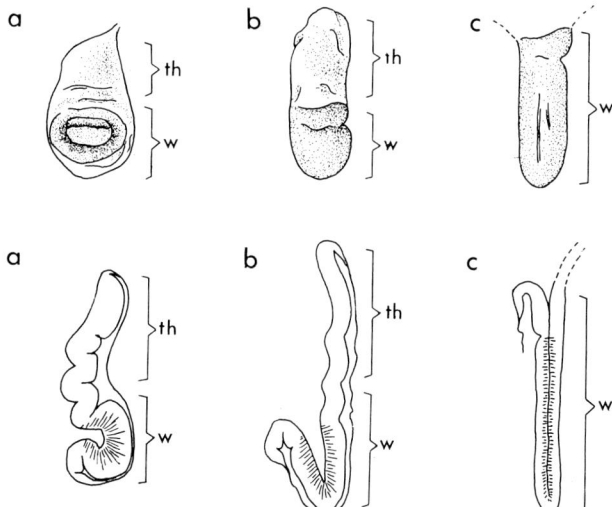

Fig. 21. Evagination of an imaginal wing disk. Evagination is pictured both in the whole disk and in a cross section though the disk. Note that the area occupied by the wing increases dramatically in size, and also that the cells flatten during the evagination process (From FRISTROM, 1969a)

Organized changes in cell shape have been detected in other systems and found to be the causes of morphogenetic movements. Such changes in cell shape have frequently been found to involve the action of intracellular contractile elements. For example, BAKER and SCHROEDER (1967) have shown that the change in cell shape associated with the invagination of the neural plate during amphibian development is caused by contractile filaments. KARFUNKEL (1971) has implicated both microtubules and microfilaments in this process. BURNSIDE and JACOBSON (1968) have suggested that contractile proteins are involved in the morphogenetic movements of the neural plate which precede invagination during the development of the newt, *Taricha torosa*.

It seems likely that similar proteins are involved in the change in cell shape in disks. Microtubules have been detected in disk cells (URSPRUNG and SCHABTACH, 1968; WEHMAN, 1969; FRISTROM, 1969b; POODRY and SCHNEIDERMAN, 1970) and are principally located along the long axes of the cells. However, the major evidence supporting the role of contractile proteins in evagination depends upon the use of inhibitors (Table 7). Colchicine, an agent which interferes with microtubular function, does not inhibit evagination. Also, POODRY and SCHNEIDERMAN (1971) have demonstrated that colcemid, at concentrations which cause complete loss of detectable microtubules, does not inhibit evagination. Vinblastin, an agent which precipitates microtubular and apparently microfiber proteins (KARFUNKEL, 1971), and also a variety of cellular proteins and components (WILSON et al., 1970) inhibits evagination. Cytochalasin B which inhibits cytokinesis but does not interfere with karyokinesis (CARTER, 1967) and therefore does not inhibit functions mediated by microtubular proteins, inhibits evagination. SPOONER and WESSELLS (1970) report that

Table 7. Inhibition of evagination of disks

Compound tested	Concentration range tested	Minimum effective concentration
Colchicine	$10^{-7}-10^{-5}$ M^a	No inhibition
Vinblastin	$10^{-4}-10^{-6}$ M	10^{-4} M
Cytochalasin B	.01–10 µg/ml	1 µg/ml
Puromycin	1–100 µg/ml	10 µg/ml

^{a} The colchicine used was found to be active against divisions of grasshopper spermatocytes over this concentration range.

Cytochalasin B acts by causing disaggregation of microfiber proteins. Microfiber proteins, while distinct from microtubular proteins, also function in the control of cell shape (CARTER, 1967; SPOONER and WESSELLS, 1970) and are probably also contractile. Therefore, the data indicate that microfiber proteins cause evagination *via* a change in shape of the disk cells. Furthermore, it is clear that Cytochalasin B does not inhibit the effect of β-ecdysone upon disk cells, but inhibits the mechanics of evagination. In order to produce complete evagination (which requires about 12 h after the hormone is added to the medium) the disks must be exposed to ecdysone only during the first 6 h. Cytochalasin B inhibits evagination after ecdysone is no longer required. When Cytochalasin B and β-ecdysone are simultaneously incubated with disks for 6 h and then both substances removed, the disks will completely evaginate (CHIHARA, personal communication). Therefore, Cytochalasin B does not inhibit the action of ecdysone upon the cells, but only the response to the hormone.

Microfiber proteins are undoubtedly present in disk cells before exposure to β-ecdysone. Thus, there may be no requirement for *de novo* synthesis of these proteins. It is, however, clear that protein synthesis is required for evagination. Puromycin inhibits the progress of evagination at any time (FRISTROM et al., 1969) and removal of the single amino acid, leucine, from ROBB's culture medium also inhibits evagination (FRISTROM, unpublished observation)[4]. Whether β-ecdysone stimulates increased synthesis of contractile proteins or of proteins which participate in the assembly of the contractile elements remains to be determined. POODRY and SCHNEIDERMAN (1971) have found that trypsin treatment of late third instar disks causes

---

4 LEE (personal communication) reports that disks dissected from larvae shortly before puparium formation are capable of evagination in *Drosophila* RINGER's (EPHRUSSI and BEADLE, 1936). This observation suggests changes in disk physiology shortly before puparium formation since the mass-isolated disks which we use, as noted above, are isolated from younger larvae.

evagination. Thus, factors reducing cellular adhesivity may also be involved along with contractile proteins in producing evagination.

In order for any force which results in evagination to be effective it must operate under carefully specified conditions so that the resulting appendage will be properly structured. These conditions are supplied in the architectural structure of the larval disks. Thus, the production of legs as narrow tubes with the segments clearly defined and wings as large pouches requires the existence of areas in the disks which control and direct the mechanical force supplied by the flattening of the cells. Thus, some cells must supply, to use SPREIJ's term (SPREIJ, 1970) the "skeleton" for the disks. These cells presumably do not flatten during evagination. The geographic localization of some of the skeletal elements in 3rd instar disks has already been determined by SPREIJ (1970) through the use of histochemistry. He has discovered the 5'-nucleotidase is mainly localized laterally at the bases of the future appendages. He points out that 5'-nucleotidase has been localized in cell membranes and may be important in maintaining the rigidity of the membranes. SPREIJ also notes that the microtubules are located along the long axes of the columnar cells and suggests that the combination of microtubules and the 5'-nucleotidase may maintain the rigidity of these cells and thus control to some degree the shape of the appendage.

VIII. Regulatory Mechanisms in Disk Metamorphosis

The major events in imaginal disk morphogenesis occur over a comparatively short period of time after an extensive period of preparation. Thus, in *D. melanogaster* the imaginal disks form almost the complete external structure of the adult insect in a 12 h period. This rapid period of morphogenesis was preceded by 96 h of larval life and 20 h of embryogenesis. Larvae have been considered by some authors to be "feeding embryos". However, consideration of the developmental processes involved in disk morphogenesis leads one to view the larval form, in so far as disk development is concerned, not as a feeding embryo, but rather as a feeding oocyte. Embryogenesis and disk metamorphosis have many characteristics in common. First, they are initiated by specific stimuli, the former by fertilization, the latter by ecdysone. Second, substantial periods of preparation precede both developmental periods. Third, once development is initiated it goes forward with specified steps in a highly coordinated manner. Fourth, and perhaps most important, both processes are initiated only after the preceding developmental stage has reached adequate maturity.

Some of the similarities between embryogenesis and disk metamorphosis are illustrated by a consideration of the development of the mutant *bobbed (bb)*. The *bb* mutant is characterized by a reduction in the number of rRNA genes present (RITOSSA, ATWOOD and SPIEGELMAN, 1966). The major characteristics of the phenotype are reduction in bristle size, etching of the abdomen and slow development. It has been thought that development is slowed throughout all stages of the life cycle of *bb* individuals (MOHAN and RITOSSA, 1970). However, careful measurements of the durations of different phases of development reveal that this is not the case. Egg maturation, as judged by the rate at which *bb* females lay eggs, is retarded (MOHAN and RITOSSA, 1971). However, embryogenesis proceeds at similar rates in wild type and a variety of *bb* stocks (STEWART, 1971). Larval development is retarded in *bb* stocks. This is demonstrated most spectacularly in $X^{bb}0$ males (containing less than 15%

of the wild type number of rRNA genes) where larval development lasts 176 h at 25° C (compared to 90 h for Ore R controls). The prepupal periods, during which metamorphosis occurs, are essentially identical in length (12.15 h for $X^{bb}0$ males; 12.05 h for Ore R males). The time required for pupal development is also similar (101 h for Ore R males; 108 h $X^{bb}0$ males). Evagination of wild type and bb disks also occurs at similar rates *in vitro*. Thus, it is not the periods of differentiation (embryogenesis and metamorphosis) which are affected by the bb genotype, but rather the periods of growth and maturation (oogenesis and larval development). These data indicate that rRNA synthesis is not a rate limiting factor during differentiation, and thus the requirements for rRNA must have already been satisfied prior to the initiation of metamorphosis. This is confirmed by the demonstration that late larval bb disks have the same content of DNA and RNA as Ore R disks.

The fact that the ribosome requirement for disk differentiation is met during the long maturation period illustrates the molecular strategy used by disks during development. It seems highly likely that other synthetic processes in disks are similarly "prepared" for metamorphosis, thereby minimizing the amount of new synthesis required during the period of rapid differentiation. One obvious way to minimize synthesis is to exert regulation at the post-tanscriptional level, i.e. at the translational level or at the level of enzyme activity. A clear precedent for this type of regulation is provided by the translational control mechanisms used during early embryogenesis (e.g. masked messenger RNA; GROSS and COUSINEAU, 1963; DENNY and TYLER, 1964; STAVY and GROSS, 1967). In view of the many similarities between embryogenesis and metamorphosis discussed above, it would not be surprising to find similar types of regulatory mechanisms operative in the metamorphosis of disks.

Regulation during disk metamorphosis of the synthesis of new proteins and enzymes at the translational rather than at the transcriptional level would obviate any requirement for extensive new RNA synthesis. However, most of the available evidence indicates that the primary action of ecdysone is on the genome of disk cells. This evidence is critically examined below.

1) Analogy to polytene cells: Ecdysone induces new puffs in polytene chromosomes (BERENDES, 1967; ASHBURNER, 1971). This direct visualization of gene action establishes that ecdysone does stimulate new transcription in *Drosophila* cells. However, all polytene tissues studied are juvenile and are generally programmed to degenerate and not to undergo rapid and extensive differentiation. It would be dangerous to assume that the mode of action of ecdysone on imaginal and juvenile tissues is necessarily the same.

2) Effect of Actinomycin D: Ecdysone-induced evagination is inhibited by Actinomycin D during the first 6 h after the addition of β-ecdysone to the culture medium (FRISTROM et al., 1969). This is the same period of time during which β-ecdysone is required in the culture medium to produce complete evagination. These observations suggest that β-ecdysone stimulates the synthesis of new RNAs needed for evagination. However, as pointed out above, Actinomycin D, by preventing the replenishment of mRNAs, also inhibits protein synthesis. Since protein synthesis is required for evagination, the inhibition by Actinomycin D may result from the indirect inhibition of protein synthesis and not from the inhibition of synthesis on new mRNAs.

3) Effect of β-ecdysone upon incorporation into RNA: β-ecdysone causes a significant increase in the incorporation of labeled precursors into RNA which cannot be accounted for by increased uptake of the precursor, increased stability of RNA or increased metabolism of the precursor. These data strongly suggest that β-ecdysone causes increased synthesis of RNA in disks.

4) Effect of β-ecdysone upon the hybridization of pulse labeled disk RNA: Although ecdysone apparently increases RNA synthesis it is not clear whether the increased synthesis involves the production of new species of RNA molecules. Ecdysone does induce changes in the hybridizing capacity of pulse-labeled RNA. This indicates that the hormone causes changes in net synthesis of classes of RNA molecules, but does not demonstrate that any new RNA species are synthesized. Data from competition experiments also fail to reveal any qualitative change in the RNA molecules synthesized in disks in the presence and absence of β-ecdysone. However, any new species of RNA transcribed from unique DNA would not be detected by the technique used.

5) Temporal sequence of events: β-ecdysone has a rapid effect upon incorporation into RNA, rRNA synthesis and the hybridyzing capacity of pulse labeled RNA. These effects are all detected within an hour after the addition of the hormone and some occur within 20—30 min (Fig. 22). The effect upon incorporation into protein is,

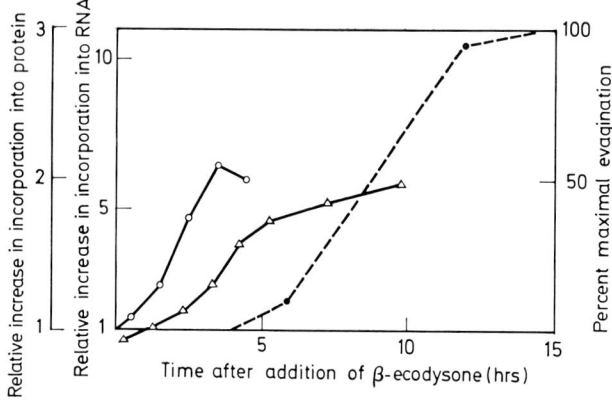

Fig. 22. Temporal sequence of biosynthetic and morphogenetic effects of β-ecdysone. Relative incorporation into RNA (—O—); protein (—△—) and time course of evagination (------) are depicted. Control disks, by definition, have relative incorporation of one (FRISTROM, unpublished)

however, comparatively slow, with a substantial effect occurring only 3—4 h after the addition of the hormone. Evagination itself is not detected until about 6 h after the addition of β-ecdysone. The rapidity of the effect upon RNA metabolism is thus a strong indication that the primary effect of ecdysone is upon transcription, while the increase in incorporation into protein is indirect. Evagination can occur in the absence of β-ecdysone and thus no direct participation of the hormone is required in the mechanical aspects of this process.

The evidence presented above indicates that the primary effect of β-ecdysone is upon transcription of RNA. This conclusion, however, does not eliminate the possibility that the hormone induces transcription of only a limited number of new genes. The possibility that ecdysone, although apparently directly affecting transcription, indirectly affects translation has also been investigated. Evidence is now accumulating which indicates that ribosomal proteins may restrict the translation of specific mRNAs. HEYWOOD (1969, 1970) has demonstrated that the translation of myosin mRNA by reticulocyte ribosomes is dependent upon the addition of protein factors from muscle ribosomes. Working with *Tenebrio* the ILANS (ILAN and ILAN, 1971) have demonstrated that proteins isolated from pupal ribosomes must be added to

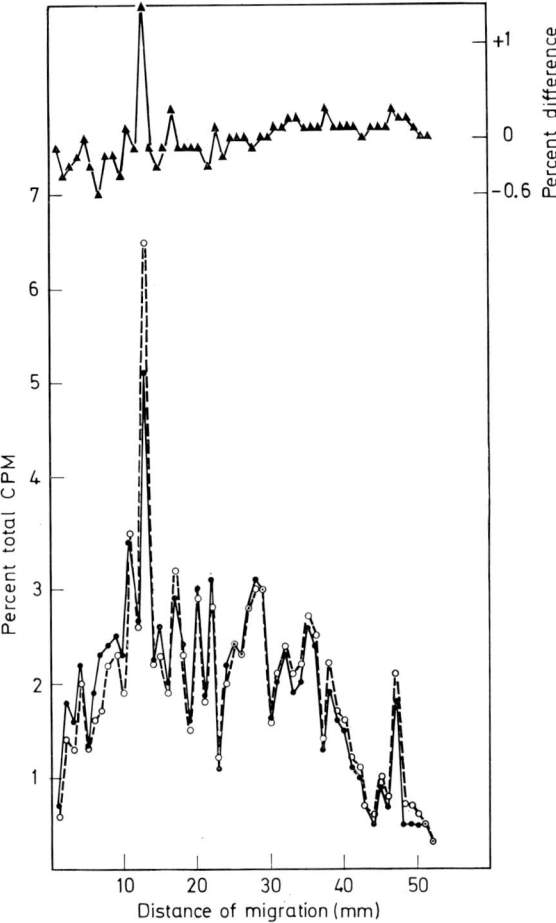

Fig. 23. Effect of β-ecdysone upon synthesis of proteins extracted by high salt from ribosomes. Disks were incubated in ROBB's culture medium at 25° C for 5 h with (------) and without (———) β-ecdysone and then respectively with ^{14}C-leucine and ^{3}H-leucine for 2 h. The ribosome pellet was extracted for 1 h with 1 M KCl and the proteins subjected to SDS acrylamide gel electrophoresis. The distribution of counts along the gel and the difference plot between ^{14}C- and ^{3}H-protein is depicted. Note that only a single section of the gel is substantially affected (SIEGEL, GREGG, and FRISTROM, unpublished)

larval ribosomes to allow binding of pupal mRNAs. In both examples the protein factors involved have been isolated by a high salt wash of the ribosomes, and therefore this fraction of ribosomal proteins contains factors which regulate the translation of specific mRNAs. The possibility has therefore been investigated that β-ecdysone induces synthesis of new proteins found in this fraction of ribosomal proteins. Following a 5 h preincubation with β-ecdysone it is possible to detect increased synthesis of a protein found in the high salt wash of disk ribosomes (Fig. 23). Although the synthesis of a protein found in this fraction of ribosomal proteins has been affected by β-ecdysone, it is yet to be established that this protein has a function in the regulation of translation. The *de novo* synthesis of a translational regulatory factor in disks in response to β-ecdysone is an attractive possibility. It provides a regulatory mechanism by which synthesis of new proteins could be initiated without requiring *de novo* synthesis of new types of mRNA following exposure to ecdysone. However, it must be emphasized that although this hypothesis is highly plausible, it is still speculative.

In summary, a general hypothesis for the induction of disk evagination by ecdysone involves the following steps:

1) Ecdysone induces the synthesis of increased amounts of RNA, but presumably only a limited number of new mRNAs. Since ecdysone has been found to enter epidermal cells (KARLSON et al., 1964) we presume it also enters disk cells and thus may directly affect transcription without the intervention of a "second message" such as cyclic AMP which has been found to mediate the action of many hormones (for review see ROBISON, BUTCHER and SUTHERLAND, 1968). Although work in BERENDES' laboratory (LEENDERS, WULLEMS and BERENDES, 1970) demonstrates that cyclic AMP potentiates the action of ecdysone on salivary glands, we have not found a similar potentiating effect with disks (CHIHARA, personal communication).

2) An increase in protein synthesis results from the increased production of RNA in disk cells. The increase in protein synthesis may result in part from the presence of a new translational factor which allows translation of pre-existing mRNAs.

3) The proteins synthesized participate in the orientation, assembly or function of microfibers and produce a change in cell shape from columnar to cuboidal epithelium.

4) The change in cell shape causes an increase in surface area and provides the mechanical force which in conjunction with pre-existing anatomical conditions results in evagination of the different disk appendages.

Note added in proof

Since completing the manuscript a paper has come to my attention (THOMSON, J. A., ROGERS, D. C., GUNSON, M. M., HORN, D. H. S.: Cytobios 6, 79—88, 1970) which shows that B-ecdysone is transported into the nucleus and cytoplasm of imaginal disks of *Calliphora stygia*.

Acknowledgements

I am indebted to many people for helpful discussions during the preparation of this manuscript. In particular I wish to thank Dr. COLIN MURPHY and Dr. JOHN MERRIAM for their helpful discussions on the use of cell lineage techniques for estimating cell number in imaginal disks, and Dr. DAVID KING for discussions on insect hormone physiology. I am extremely grateful to Dr. CHARLES LAIRD for his patient instruction in the techniques of RNA/DNA hybridization. To my wife, who carefully read the manuscript and gave it what clarity it possesses (frequently over my own opposition!) I owe special appreciation. I also gratefully acknowledge the expert technical assistance of Mrs. ODESSA EUGENE and Miss LEE GREGG.

References

Agui, N., Yagi, S., Fukaya, M.: Effects of ecdysterone on the *in vitro* development of wing discs of rice stem borer, *Chilo suppressalis*, Appl. Ent. Zool. **4**(3), 158—159 (1969).

Ashburner, M.: Induction of puffs in polytene chromosomes of *in vitro* cultured salivary glands of *Drosophila melanogaster* by ecdysone and ecdysone analogues. Nature (Lond.) **230**, 222—223 (1971).

Auerbach, C.: The development of the legs, wings and halteres in wildtype and some mutant strains of *Drosophila melanogaster*. Trans. roy. Soc. Edinburgh **58**, 787—815 (1936).

Baker, P. C., Schroeder, T. E.: Cytoplasmic filaments and morphogenetic movement in the amphibian neural tube. Develop. Biol. **15**, 432—450 (1967).

Becker, H. J.: Über Röntgenmosaikflecken und Defektmutationen am Auge von *Drosophila* und die Entwicklungsphysiologie des Auges. Z. indukt. Abstamm- u. Vererb.-L. **88**, 333—373 (1957).

Berendes, H. D.: The hormone ecdysone as effector of specific changes in the pattern of gene activities in *Drosophila hydei*. Chromosoma **22**, 274—293 (1967).

Bodenstein, D.: Growth regulation of transplanted eye and leg discs in *Drosophila*. J. exp. Zool. **84**, 23—39 (1940).

Britten, R. J., Kohne, D. E.: Repeated sequences in DNA. Science **161**, 529—540 (1968).

Bryant, P. J.: Cell lineage relationships in the imaginal wing disk of *Drosophila melanogaster*. Develop. Biol. **22**, 389—411 (1970).

Bryant, P. J., Schneiderman, H. A.: Cell lineage, growth, and determination in the imaginal leg discs of *Drosophila melanogaster*. Develop. Biol. **20**, 263—290 (1969).

Burdette, W. J.: Changes in titer of ecdysone in *Bombyx mori* during metamorphosis. Science **135**, 432 (1962).

Burnside, M. B., Jacobson, A. G.: Analysis of morphogenetic movements in the neural plate of the newt *Taricha torosa*. Develop. Biol. **18**, 537—552 (1968).

Carter, S. B.: Effects of cytochalasins on mammalian cells. Nature (Lond.) **213**, 261—264 (1967).

Chiarodo, A. J., Denys, F. R.: Fine structural features of developing leg discs of the blowfly, *Sarcophaga bulleta*. J. Morph. **126**, 349—364 (1967).

Chihara, C., Petri, W., Fristrom, J., King, D.: The assay of ecdysones and juvenile hormones in *Drosophila* imaginal disks *in vitro*. J. Insect Physiol. (in press).

Darnell, J. E.: Ribonucleic acids from animal cells. Bact. Rev. **32**, 262—290 (1968).

Denny, P. C., Tyler, A.: Activation of protein biosynthesis in non-nucleate fragments of sea urchin eggs. Biochem. Biophys. Res. Commun. **14**, 245—249 (1964).

Edstrom, J. E., Daneholt, B.: Sedimentation properties of the newly synthesized RNA from isolated nuclear components of *Chironomus tentans* salivary gland cells. J. molec. Biol. **28**, 331—343 (1967).

El Shatoury, H. H., Waddington, C. H.: The development of gastric tumours in *Drosophila* larvae. J. Embryol. exp. Morphol. **5**, 143—152 (1957).

Ephrussi, B., Beadle, G. W.: A technique of transplantation for *Drosophila*. Amer. Naturalist **70**, 218—225 (1936).

Feir, D., Winkler, G.: Ecdysone titres in the last larva and adult stages of the milkweed bug. J. Insect. Physiol. **15**, 899—904 (1969).

Fristrom, D.: Cellular degeneration in the production of some mutant phenotypes in *Drosophila melanogaster*. Molec. Gen. Genetics **103**, 363—379 (1969a).

Fristrom, D.: Ultrastructure of developing wildtype and mutant strains of *Drosophila melanogaster*. Ph. D. Thesis, University of California, Berkeley (1969b).

Fristrom, J. W.: Development of the morphological mutant cryptocephal of *Drosophila melanogaster*. Genetics **52**, 297—318 (1965).

Fristrom, J. W.: Hexosamine metabolism of imaginal disks of *Drosophila melanogaster*. J. Insect Physiol. **14**, 729—740 (1968).

Fristrom, J. W.: The developmental biology of *Drosophila*. Ann. Rev. Genetics **4**, 325—345 (1970).

Fristrom, J. W., Brothers, L., Mancebo, V., Stewart, D.: Aspects of RNA and protein synthesis in imaginal discs of *Drosophila melanogaster*. Molec. Gen. Genet. **102**, 1—14 (1968).

Fristrom, J. W., Heinze, W.: The preparative isolation of imaginal discs. Drosophila Inf. Serv. **43**, 186 (1968).

Fristrom, J. W., Knowles, B. B.: Studies on protein synthesis in imaginal discs of *Drosophila melanogaster*. Exp. Cell Res. **47**, 97—107 (1967).

Fristrom, J. W., Mitchell, H. K.: The preparative isolation if imaginal discs from larvae of *Drosophila melanogaster*. J. Cell Biol. **27**, 445—448 (1965).

Fristrom, J. W., Raikow, R., Petri, W., Stewart, D.: *In vitro* evagination and RNA syn-synthesis in imaginal discs of *Drosophila melanogaster*. In: Hanly, E. W., (Ed.): Park City Symposium on Problems in Biology, pp. 381—401. Salt Lake City: Univ. Utah Press 1969.

Gall, J. G., Cohen, E. H., Polan, M. L.: Repetitive DNA sequences in *Drosophila*. Chromosoma **33**, 319—344 (1971).

Garcia-Bellido, A., Merriam, J. R.: Parameters of the wing imaginal disc development of *Drosophila melanogaster*. Develop. Biol. **24**, 61—87 (1971).

Gateff, E., Schneiderman, H. A.: Neoplasms in mutant and cultured wildtype tissues of *Drosophila*. National Cancer Inst. Monograph **31**, 365—397 (1969).

Gehring, W.: Bildung eines vollständigen Mittelbeines mit Sternopleura in der Antennenregion bei der Mutante Nasobemia (Ns) von *Drosophila melanogaster*. Arch. Julius Klaus-Stiftung **41**, 44—54 (1966).

Giles, C. H., Hassan, A. S. A., Laidlaw, M., Subramanian, R. V. R.: Adsorption at organic surfaces. III. Some observations on the constitution of chitin and on its adsorption of inorganic and organic acids from aqueous solution. J. Soc. Dy. Col. **74**, 647—654 (1958).

Glaser, L., Brown, D. H.: The synthesis of chitin in cell-free extracts of *Neurospora crassa*. J. biol. Chem. **228**, 729—742 (1957).

Gross, P. R., Cousineau, G. H.: Effects of actinomycin D on macromolecular synthesis and early development in sea urchin eggs in response to fertilization. Exp. Cell Res. **25**, 405—417 (1963).

Heywood, S. M.: Synthesis of myosin on heterologous ribosomes. Cold Spr. Harb. Symp. quant. Biol. **34**, 799—803 (1969).

Heywood, S. M.: Specificity of mRNA binding factor in eukaryotes. Proc. nat. Acad. Sci. (Wash.) **67**, 1782—1788 (1970).

Ilan, J., Ilan, J.: Stage-specific initiation factors for protein synthesis during insect development. Develop. Biol. **25**, 280—292 (1971).

Kafatos, Fotis C.: Cocoonase synthesis: cellular differentiation in developing silk moths. In: Hanly, E. W., (Ed.): Park City Symposium on Problems in Biology, pp. 111—140. Salt Lake City: Univ. Utah Press 1969.

Kaplanis, J. N., Thompson, M. J., Yamamoto, R. I., Robbins, W. E., Louloudes, S. J.: Ecdysones from the pupa of the tobacco hornworm *Manduca sexta* (Johannson). Steroids **8**, 605—623 (1966).

Karfunkel, P.: The role of microtubules and microfilaments in neuralution in *Xenopus*. Develop. Biol. **25**, 30—56 (1971).

Karlson, P., Sekeris, C. E., Maurer, R.: Zum Wirkungsmechanismus der Hormone I. Verteilung von tritium-markierten Ecdyson in Larven von *Calliphora erythrocephala*. Z. physiol. Chem. **336**, 100—106 (1964).

Krishnakumaran, A., Berry, S. J., Oberlander, H., Schneiderman, H. A.: Nucleic acid synthesis during insect development-II. Control of DNA synthesis in the *Cecropia* silkworm and other saturniid moths. J. Insect Physiol. **13**, 1—57 (1967).

Kurnick, N. B., Herskowitz, I. H.: The estimation of polyteny in *Drosophila* salivary gland nuclei based on determination of DNA content. J. Cell. Comp. Physiol. **39**, 281—299 (1952).

Laird, C. D.: Chromatid structure: Relationship between DNA content and nucleotide sequence diversity. Chromosoma **32**, 378—406 (1971).

Law, J. H., Yuan, C., Williams, C. M.: Synthesis of a material with high juvenile hormone activity. Proc. nat. Acad. Sci. (Wash.) **55**, 576—578 (1966).

Leenders, H. J., Wullems, G. J., Berendes, H. D.: Competitive interaction of adenosine 3′,5′-monophosphate on gene activation by ecdysterone. Exp. Cell. Res. **63**, 159—164 (1970).

Lewis, E. B.: Genetic control and regulation of developmental pathways. In: Locke, M., (Ed.): The Role of Chromosomes in Development, pp. 231—252. New York-London: Academic Press 1964.

Locke, M.: The structure and formation of the cuticulin layer in the epicuticle of an insect *Calpodes ethlius* (Lepidoptera, Hespeiidae). J. Morph. **118**, 461—494 (1966).

Locke, M.: The structure of an epidermal cell during the development of the protein epicuticle and the uptake of molting fluid in an insect. J. Morph. **127**, 7—40 (1969).

Mandaron, P.: Développement *in vitro* des disques imaginaux de la *Drosophile*. Aspects morphologiques et histologiques. Develop. Biol. **22**, 298—320 (1970).

Mandaron, P.: Sur le mécanisme de l-évagination des disques imaginaux chez la *Drosophile*. Develop. Biol. **25**, 581—605 (1971).

Meyer, A. S., Hanzmann, E., Schneiderman, H. A.: The isolation and identification of the two juvenile hormones from the Cecropia silk moth. Arch. Biochem. Biophys. **137**, 190—213 (1970).

Mohan, J., Ritossa, F. M.: Regulation of ribosomal RNA synthesis and its bearing on the *bobbed* phenotype of *Drosophila melanogaster*. Develop. Biol. **22**, 495—512 (1970).

Moriyama, H., Nakanishi, K., King, D. S., Okauchi, T., Siddal, J. B., Hafferl, W.: On the origin and metabolic fate of α-ecdysone in insects. Gen. Comp. Endocrin. **15**, 80—87 (1970).

Munro, H., Flick, A.: The determination of nucleic acids. In: Glick, D., (Ed.): Methods of Biochemical Analysis, Vol. 14, pp. 113—176. New York: Wiley; Interscience 1966.

Oberlander, H.: Effects of ecdysone, ecdysterone, and inokosterone on the *in vitro* initiation of metamorphosis of wing disks of *Galleria mellonella*. J. Insect. Physiol. **15**, 297—304 (1969a).

Oberlander, H.: Ecdysone and DNA synthesis in cultured wing disks of the greater wax moth, *Galleria mellonella*. J. Insect Physiol. **15**, 1803—1806 (1969b).

Oberlander, H., Fulco, L.: Growth and partial metamorphosis of imaginal disks of the greather wax moth, *Galleria mellonella*, *in vitro*. Nature (Lond.) **216**, 1140—1141 (1967).

Ohtaki, T., Milkman, R. D., Williams, C. M.: Ecdysone and ecdysone analogues. Their assay on the fleshfly *Sarcophaga peregrina*. Proc. nat. Acad. Sci. (Wash.) **58**, 981—984 (1967).

Ouweneel, W. J.: Morphology and development of loboid ophthalmoptera, a homoeotic mutant of *Drosophila melanogaster*. Wilhelm Roux' Arch. Entwickl.-Mech. Org. (1969a).

Ouweneel, W. J.: Influence of environmental factors on the homoeotic effect of loboid ophthalmoptera in *Drosophila melanogaster*. Wilhelm Roux' Arch. Entwickl.-Mech. Org. **164**, 15—36 (1969b).

Patel, N., Madhavan, K.: Effects of hormones on RNA and protein synthesis in the imaginal wing disks of the ricini silkworm. J. Insect. Physiol. **15**, 2141—2150 (1969).

Petri, W. H., Fristrom, J. W., Stewart, D. J., Hanly, E. W.: The *in vitro* synthesis and characteristics of ribosomal RNA in imaginal discs of *Drosophila melanogaster*. Molec. Gen. Genetics **110**, 245—262 (1971).

Poodry, C. A., Schneiderman, H. A.: The ultrastructure of the developing leg of *Drosophila melanogaster*. Wilhelm Roux' Arch. Entwickl.-Mech. Org. **166**, 1—44 (1970).

Poodry, C. A., Schneiderman, H. A.: Intercellular adhesivity and pupal morphogenesis in *Drosophila melanogaster*. Wilhelm Roux' Arch. Entwickl.-Mech. Org. **168**, 1—9 (1971).

Postlethwait, J. H., Schneiderman, H. A.: Induction of metamorphosis by ecdysone analogues: *Drosophila* imaginal discs cultured *in vivo*. Biol. Bull **138**, 47—55 (1970).

Raikow, R., Fristrom, J. W.: Effects of β-ecdysone on RNA metabolism of imaginal discs of *Drosophila melanogaster*. J. Insect Physiol. **17**, 1599—1614 (1971).

Richards, A. G.: The integument of Arthropods, pp. 1—411. Minneapolis: Univ. Minnesota Press 1951.

Ritossa, F. M., Atwood, K. C., Spiegelman, S.: A molecular explanation of the *bobbed* mutants of *Drosophila* as partial deficiencies of "ribosomal" DNA. Genetics **54**, 819—834 (1966).

Robb, J. A.: Maintenance of imaginal discs of *Drosophila melanogaster* in chemically defined media. J. Cell Biol. **41**, 876—884 (1969).

Robison, G. A., Butcher, R. W., Sutherland, E. W.: Cyclic AMP. Annual Rev. Biochem. **37**, 149—174 (1968).

Robson, E. A.: The cuticle of *Peripatopsis moseleyi*. Quart. J. Micr. Sci. **105**, 281—294 (1964).

Röller, H., Dahm, K. H., Sweeley, C. C., Trost, B. M.: The structure of the juvenile hormone. Angew. Chem. Int. Ed. Engl. **6**, 179—180 (1967).

Schneider, I.: Differentiation of larval *Drosophila* eye-antennal discs *in vitro*. J. exp. Zool. **156**, 91—104 (1964).

Sehnal, F.: Influence of the corpus allatum on the development of internal organs in *Galleria mellonella* L. J. Insect Physiol. **14**, 73—85 (1968).

Sengel, P., Mandaron, P.: Aspects morphologiques du développement *in vitro* des disques imaginaux de la *Drosophila*. C. R. Acad. Sci. Paris **268**, 405—407 (1969).

Shaaya, E., Karlson, P.: Der Ecdysontiter während der Insektenentwicklung-IV. Die Entwicklung der Lepidopteren *Bombyx mori* und *Cerura vinula* L. Develop. Biol. **11**, 424—432 (1965a).

Shaaya, E., Karlson, P.: Der Ecdysontiter während der Insektenentwicklung-II. Die postembryonale Entwicklung der Schmeißfliege, *Calliphora erythrocephala*. J. Insect Physiol. **11**, 65—69 (1965b).

Spooner, B. S., Wessells, N. K.: Effects of cytochalasin B upon microfilaments involved in morphogenesis of salivary epithelium. Proc. nat. Acad. Sci. **66**, 360—364 (1970).

Spreij, Th. E.: Cell death during the development of the imaginal disks of *Calliphora erythrocephala*. Neth. J. Zool. **21**, 221—264 (1971).

Spreij, Th. E.: Localization of 5'-nucleotidase and its possible significance in some of the imaginal disks of *Calliphora erythrocephala*. Neth. J. Zool. **20**, 419—432 (1970).

Stavy, L., Gross, P. R.: The protein-synthetic lesion in unfertilized eggs. Proc. nat. Acad. Sci. (Wash.) **57**, 735—742 (1967).

Stewart, D. J.: RNA synthesis in imaginal discs of a *bobbed* mutant of *Drosophila melanogaster*. Ph. D. Thesis, University of California, Berkeley (1971).

Stewart, M., Murphy, C., Fristrom, J. W.: The recovery and preliminary characterization of x chromosome mutants affecting imaginal discs of *Drosophila melanogaster*. Develop. Biol. **27**, 71—83 (1972).

Stumpf, H.: Die Richtungen die Teilungsspindeln auf dem Puppenflügel von *Drosophila* in Verlaufe der Mitosenperiode. Biol. Zbl. **75**, 17—27 (1956).

Suzuki, D. T.: Temperature sensitive mutations in *Drosophila melanogaster*. Science **170**, 695—706 (1970).

Tartof, K. D., Perry, R. P.: The 5S RNA genes of *Drosophila melanogaster*. J. Molec. Biol. **51**, 171—183 (1970).

Tata, J. R.: Hormonal regulation of growth and protein synthesis. Nature (Lond.) **219**, 331—337 (1968).

Ursprung, H., Schabtach, E.: The fine structure of the male *Drosophila* genital disk during late larval and early pupal development. Wilhelm Roux' Arch. Entwickl.-Mech. Org. **160**, 243—254 (1968).

Ursprung, H., Sofer, W. H., Burroughs, N.: Ontogeny and tissue distribution of alcohol dehydrogenase in *Drosophila melanogaster*. Wilhelm Roux' Arch. Entwickl.-Mech. Org. **164**, 201—208 (1970).

Wehman, H. J.: Fine structure of *Drosophila* wing imaginal discs during early stages of metamorphosis. Wilhelm Roux' Arch. Entwickl.-Mech. Org. **163**, 375—390 (1970).

Wilson, L., Bryan, J., Ruby, A., Mazia, D.: Precipitation of proteins by vinblastine and calcium ions. Proc. nat. Acad. Sci. (Wash.) **66**, 807—814 (1970).

Wyatt, G. R.: Biochemistry of insect metamorphosis. In: Etkin, E., Gilbert, L. I., (Eds.): Metamorphosis, pp. 143—184. New York: Appleton Century Crofts 1968.

Zweidler, A., Cohen, L. H.: Large scale isolation and fractionation of organs of *Drosophila melanogaster* larvae. J. Cell Biol. **51**, 240—248 (1971).

The Hormonal Control of Development of Imaginal Disks

HERBERT OBERLANDER

*Market Quality Research Division, Agricultural Research Service,
United States Department of Agriculture, Gainesville, Florida*

I. Introduction

Until recently the endocrine control of imaginal disk development had escaped detailed attention. Insect endocrinologists have largely bypassed the imaginal disks by focusing on external changes in hemimetabolous insects, and on the pupal-adult transformation in holometabolous insects. While we have learned a great deal about the control of molting and metamorphosis by ecdysone and juvenile hormone, we know little about how these hormones control the life of an imaginal disk. In the absence of juvenile hormone, ecdysone causes metamorphosis, and hence the differentiation of the imaginal disks. It need not follow that the disks are direct targets of ecdysone. Nor is it at all apparent how the proliferation of young larval disks is controlled. Our understanding of how juvenile hormone prevents the differentiation of such disks is exceedingly meager. However, progress is being made, particularly with *in vitro* studies.

The actions of insect hormones on imaginal disks have now been examined in several species of insects. Most prominent among these are: *Calliphora erythrocephala* (Diptera), *Drosophila melanogaster* (Diptera), *Ephestia kühniella* (Lepidoptera), and *Galleria mellonella* (Lepidoptera). These insects have been investigated extensively by students of insect development. We know more about the genetics of *Drosophila* than any other metazoan, and this has made *Drosophila* the organism of choice for many investigations. *Calliphora* has been important in the isolation and identification of ecdysone, while both *Ephestia* and *Galleria* have been widely employed in a variety of endocrine studies. Section II of this chapter will summarize current views of the role of brain hormone, ecdysone and juvenile hormone in the metamorphosis of these and related insects. The reader may wish to consult more detailed reviews of insect endocrinology (e.g., BERKOFF, 1969; BOWERS, 1971; HORN, 1971; JOLY, 1968; NOVAK, 1966; ROBBINS et al., 1971; WIGGLESWORTH, 1954, 1964). None of these reviews, however, focuses on the hormonal control of the imaginal disks. Subsequent sections of this chapter will deal with the actions of insect hormones on imaginal disks both *in vivo* and *in vitro*.

II. Hormonal Control of Molting and Metamorphosis

A. Brain Hormone

In a series of classic ligature and transplantation experiments with the gypsy moth, *Porthetria dispar*, KOPEC (1917, 1922) demonstrated that the brain was necessary for metamorphosis in insects. KOPEČ concluded that "the brain accordingly would have to play the 'role' of an organ with internal secretion". The neurosecretory cells of the brain which control molting were later localized in the pars intercerebralis of a variety of insects (WIGGLESWORTH, 1939, 1940; NAYAR, 1956; VAN DER KLOOT, 1961; GIRARDIE, 1964).

Two sorts of molecules with brain hormone activity have been isolated. KOBAYASHI et al. (1962) found that cholesterol isolated from *Bombyx mori* could induce molting. However, KRISHNAKUMARAN and SCHNEIDERMAN (1965) have suggested that the cholesterol apparently had brain hormone activity because it mimicked juvenile hormone or ecdysone itself. The view that brain hormone is a polypeptide has received wider support. ICHIKAWA and ISHIZAKI (1963) and ISHIZAKI and ICHIKAWA (1967) isolated a protein form *Bombyx mori* which had brain hormone activity. Their findings were confirmed by KOBAYASHI and YANAZAKI (1966) working with *Bombyx*, and by GERSCH and STURZEBECHER (1968), who used *Periplaneta*. Still other investigations suggest that brain hormone is a mucopolysaccharide or a mucoprotein (WILLIAMS, 1969). To date brain hormone has not been isolated from any Dipteran insect, or from any of the species of Lepidoptera which have been employed in the study of development of imaginal disks.

B. Molting Hormone

Although the prothoracic glands were first described in 1762 by LYONET, the role of these glands in molting was not established until FUKUDA (1940) demonstrated that silkworm larvae ligatured behind the prothorax would molt only when prothoracic glands were implanted posterior to the ligature. WILLIAMS (1947) uncovered the relationship between the brain and prothoracic glands. He showed that isolated pupal abdomens of the Cecropia silkworm did not molt if either active brains or inactive prothoracic glands were implanted. However, the abdomens metamorphosed if inactive prothoracic glands *plus active brains* were implanted.

In 1864 WEISMANN described the Dipteran counterpart of the prothoracic glands which forms a part of the ring gland. Its endocrine function was firmly established by HADORN in the course of investigating developmental mutants in *Drosophila* (HADORN, 1937; HADORN and NEEL, 1938). Lethal giant larvae (l(2)gl) is a mutant which forms a puparium only after a long delay. This genetic defect was overcome when a ring gland from a normal mature larva was transplanted into a homozygous lethal larva of the third instar. Puparium formation occurred a few hours after transplantation of the normal ring gland. Since the imaginal disks had already degenerated no development beyond puparium formation was possible. This important work was corroborated by BURTT (1938) and VOGT (1942b) who prevented pupation of larval *Drosophila* by extirpating the ring glands. In addition to confirming these early experiments POSSOMPÈS (1953) made the interesting observation that the ring glands of *Calliphora* were considerably more effective in causing pupation if they were implanted with the brain intact than with the connections severed. Thus, it appears that in Diptera the brain may control the ring gland directly via nerves rather than by releasing its hormone into the blood.

It is difficult to reconcile the various experiments reported above with the observations of PIEPHO (1948) and CHADWICK (1956) who found that the extirpation of the prothoracic glands in *Galleria* and *Periplaneta* did not prevent molting. A related result was obtained by ICHIKAWA and NISHIITSUTSUJI-UWO (1960) who reported that the brain or corpora allata of the eri-silkworm would cause molting when implanted into isolated pupal abdomens lacking prothoracic glands. These observations remain unexplained at present.

Although several ecdysones have been isolated, identified and synthesized it has never been proven that the prothoracic glands actually synthesize and release these molecules. α-ecdysone and 20-hydroxyecdysone (β-ecdysone, crustecdysone, ecdysterone) have been found in several species of *Calliphora* (KARLSON, 1956; GALBRAITH et al., 1969), and in a number of Lepidopteran insects (BUTENANDT and KARLSON, 1954; HOFFMEISTER, 1966; KAPLANIS et al., 1966). In addition 20,26-dihydroxyecdysone has been isolated in *Manduca sexta* (THOMPSON et al., 1967). The structures of both α-ecdysone and 20-hydroxyecdysone have been proven by synthesis (FURLENMEIER et al., 1965; SIDDALL et al., 1966).

We know very little about the metabolism of the ecdysones. There is good evidence that α-ecdysone can be converted to 20-hydroxyecdysone in *Calliphora* (KING and SIDDALL, 1969), *Bombyx mori* (MORIYAMA et al., 1970) and in *Antheraea polyphemus* (CHERBAS and CHERBAS, 1970). Further changes in the ecdysone molecule may result in its inactivation. A rapid inactivation of α-ecdysone by *Sarcophaga peregrina* was reported by OHTAKI et al. (1968). The enzymes which accomplish a similar inactivation of ecdysone have been localized in the fat body of *Calliphora erythrocephala* (KARLSON and BODE, 1969). It would be of some interest to learn whether the imaginal disks themselves are capable of modifying ecdysone.

C. Juvenile Hormone

The role of the corpora allata in maintaining larval development was elucidated by WIGGLESWORTH (1935, 1936, 1940), who designated the corpora allata hormone, "juvenile hormone". WILLIAMS (1961) provided conclusive evidence that the corpora allata were active throughout the third and fourth larval instars in the Cecropia silkworm, and were inactive in pupae. The actual juvenile hormone titer in Saturniid insects was measured by GILBERT and SCHNEIDERMAN (1961), who found that juvenile hormone titer was high during larval life and very low in pupae.

In the Diptera the counterpart of the corpora allata are the medial cells of the ring gland (VOGT, 1943a). A juvenilizing role for these cells was noted when their removal resulted in accelerated development of the eye disks (VOGT, 1943b). The lateral cells of the ring gland were found to be the source of molting hormone activity. When imaginal disks were implanted into the posterior portion of ligated *Drosophila* larvae they differentiated more quickly when the lateral cells of the ring gland were implanted than when the entire ring gland was used. Presumably this was because of the juvenilizing activity of the medial cells of the ring gland. Adult ring glands in which the lateral cells have degenerated prevented metamorphosis when they were transplanted into larvae of *Drosophila* (VOGT, 1946) or *Calliphora* (POSSOMPÈS, 1953).

The first extract to have juvenile hormone activity was prepared by WILLIAMS (1956) from the abdomens of adult male *Cecropia*. The extract was purified 50000 fold by WILLIAMS and LAW (1965), and after further purification was identified by RÖLLER et al. (1965, 1967) as methyl-trans, trans, cis 10 epoxy-7 ethyl 3—11 dimethyl 2,6 tridecadienoate. A second juvenile hormone was later isolated from *Cecropia* by MEYER et al. (1970). The structure of these molecules was confirmed by synthesis (JOHNSON et al., 1969; DAHM et al., 1967). It has now been demonstrated conclusively that *Cecropia* juvenile hormone is actually released by the corpora allata. RÖLLER and DAHM (1970) cultured corpora allata *in vitro*, and found that juvenile hormone accumulated in the medium.

Before 1968 there were no reports of any compounds having juvenilizing effects in the higher Diptera. Whether the Diptera possessed a unique juvenile hormone was a matter of speculation. While juvenile hormone has yet to be isolated from the Diptera, we now know that *Cecropia* juvenile hormone affects the metamorphosis of flies. SRIVASTAVA and GILBERT (1968, 1969) found that when juvenile hormone was injected into larvae of *Sarcophaga bullata* either puparium formation was prevented or development was arrested at the third day of pupal-adult development. When the hormone was applied topically to the abdomens of young pupae a second pupal cuticle was formed.

III. Imaginal Disks as Targets of Ecdysone and Juvenile Hormone
A. In vivo Experiments

We have known for three decades that the differentiation of imaginal disks at metamorphosis is under hormonal control. VOGT (1942a, 1942b, 1943b) transplanted eye-antennal disks into isolated larval abdomens of *Drosophila*, and observed that they metamorphosed only if ring glands taken from mature larvae were also transplanted. Similar results were obtained by BODENSTEIN (1943) when he transplanted disks into adult male *Drosophila melanogaster* or *Drosophila virilis*. The transplanted disks metamorphosed only if larval ring glands were simultaneously injected into the adult hosts. Surprisingly, the eye disks differentiated after two weeks in an adult female host (BODENSTEIN, 1943), producing ommochrome pigments (SCHLÄPFER, 1963). A similar situation was reported by NÖTHIGER and OBERLANDER (1967), who observed that male genital disks transplanted into adult *Drosophila melanogaster* produced rhythmically pulsating regions. Since the lateral cells of the ring gland, the counterpart of the prothoracic glands, degenerate at metamorphosis it seemed possible that an agent other than ecdysone was stimulating limited metamorphosis in the transplanted genital disks. However, NÖTHIGER and OBERLANDER (1967) obtained evidence that led them to suggest that residual ecdysone in the adults was responsible for the differentiation of pulsating genital disks. They found that the percentage of pulsating disks was higher in newly emerged female hosts than it was in older hosts. Furthermore, the response in older hosts could be increased markedly with the addition of ring glands from mature larvae. As in other experiments in which an entire ring gland is transplanted one cannot be certain whether the effect observed is due to only the lateral cells (prothoracic glands) or to the corpora allata and/or corpora cardiaca portions as well. This ambiguity was eliminated when it was found that the effect of the transplanted ring glands could be duplicated by the injection of two doses of α-ecdysone for a total of 1.2 μg hormone/gm insect (FALTUS and OBERLANDER, 1970). By contrast it took 1200 μg/gm of 20-hydroxyecdysone to induce complete metamorphosis in leg disks transplanted to an adult host (POSTLETHWAIT and SCHNEIDERMAN, 1970). Thus, it may be that only a partial metamorphosis can be attained by disks transplanted into adult hosts which still contain residual ecdysone. While we lack data on ecdysone titer in *Drosophila* adults, we do have information on several other insects. Ecdysone was not detected in adults of *Calliphora erythrocephala* (SHAAYA and KARLSON, 1965a), but was found in adults of *Bombyx mori* (SHAAYA and KARLSON, 1965b) and the milkweed bug (FEIR and WINKLER, 1969).

While there is little evidence for direct effects of juvenile hormone on imaginal disks, one dramatic instance is provided by SEHNAL (1968). Corpora allata were implanted into *Galleria* larvae on the third day of the last larval instar. An additional larval molt was produced in which the wing disks had regressed to the form characteristic of the first day of the last larval instar. SEHNAL concluded that at least the early portion of metamorphosis of the disks is sensitive to juvenile hormone.

B. In vitro Experiments

1. *Morphological Effects*

Although the experiments described in the preceding section provide convincing evidence that the metamorphosis of imaginal disks is dependent upon ecdysone, they do not distinguish between direct and indirect effects of the hormone. Direct effects of ecdysone on imaginal disks can best be demonstrated with *in vitro* experiments. Investigations of the responses of imaginal disks to hormones *in vitro* can be viewed as having gone through three phases: a) Culture of disks with brain-ring gland complexes; b) culture of disks with crude extracts with molting hormone activity; c) and culture of disks with crystalline ecdysone.

There have been a number of reports of metamorphosis of eye-antennal disks cultured with brain-ring gland complexes *in vitro* (FUGIO, 1962; GOTTSCHEWSKI, 1960; HORIKAWA, 1960; KURODA and YAMAGUCHI, 1956; SCHNEIDER, 1964). The reader is referred to a recent review by MARKS (1970) for a detailed discussion of these early papers. In considering these papers it is important to recall that stimulation of metamorphosis by the entire brain-ring gland complex does not permit a determination of which hormones are involved. There is the possibility of direct contributions from the brain, as well as the corpora allata, corpora cardiaca and prothoracic gland portions of the ring gland.

A first step in demonstrating an effect of a hormone on cultured disks was made by HORIKAWA and SUGAHARA (1960). These workers showed that delay or inhibition of metamorphosis in larvae of *Drosophila melanogaster* irradiated with varying doses of X-rays was the result of radiation damage to the brain-ring gland complex. Eye disks irradiated with doses of X-rays up to 15 Kr and cultured with 10 un-irradiated brain-ring gland complexes metamorphosed perfectly well. However, ring gland-brain complexes irradiated with only 3 Kr failed to stimulate differentiation of un-irradiated disks. The authors go on to report that "growth and differentiation" of the eye disks occurred in the presence of crude ecdysone extract of *Bombyx dupae*. Their work was confirmed by BURDETTE et al. (1968), who found that pigment deposition in the cultured eye disks was enhanced by the "ecdysone" extracts, but that improved development was obtained when the disks were cultured with entire brain-ring gland complexes.

OBERLANDER and FULCO (1967) were the first to report that crystalline alpha-ecdysone (3 µg/ml medium) induced the metamorphosis of disks cultured in a chemically defined medium. Wing disks from the last larval instar of *Galleria* were cultured in GRACE's medium without haemolymph, and responded to the ecdysone with partial metamorphosis. Subsequently, it was reported that alpha-ecdysone also stimulated metamorphosis of various *Drosophila* imaginal disks cultured *in vitro* (SENGEL and MANDARON, 1969; MANDARON, 1970); and that crystalline 20-hydroxy-ecdysone induced metamorphosis of cultured wing disks of *Galleria* (OBERLANDER, 1969a), and of the rice stem borer, *Chilo suppressalis* (AGUI et al., 1969).

The degree of morphogenesis induced by ecdysone *in vitro* varies depending on the experimental organism. In cultured wing disks of either *Galleria* (OBERLANDER, 1969a) or *Chilo* (AGUI et al., 1969) ecdysone induces tracheal migration into the lacunae, and the disks elongate. In *Chilo* a definite cuticle is formed in response to 20-hydroxyecdysone. In the cultured *Galleria* disks tritiated-glucosamine is incorporated into a chitin-like material in response to either α-ecdysone or 20-hydroxyecdysone (OBERLANDER, unpublished observations).

Tracheal migration in the cultured wing disks is one of the most sensitive and dependable responses to α-ecdysone. How does ecdysone cause these cells to move? There is considerable evidence now that microtubules are associated with movement in a variety of systems (POCHON-MASSON, 1967). The hypothesis that properly oriented microtubules are necessary for tracheal migration was tested by HASSKARL, OBERLANDER and STEPHENS (in preparation). If tracheal migration in *Galleria* disks cultured *in vitro* is dependent upon microtubules then agents which prevent or distort microtubule formation should also interfere with tracheal migration. Two such agents are colchicine and vinblastin (WHITE, 1968). Wing disks from the last larval instar of *Galleria* were cultured in GRACE's medium without haemolymph in the presence of 3 µg/ml α-ecdysone. When such disks were treated with 10^{-8} M vinblastin for 48 h tracheal migration was inhibited in 100% of the disks. A similar treatment with 10^{-3} M colchicine inhibited tracheal migration in 90% of the disks. The treated disks looked healthy, and in fact evagination took place as usual in response to α-ecdysone.

Ultrastructural analysis revealed that microtubules were present in the tracheal and tracheolar cells. The long axis of these microtubules was parallel to the long axis of the tracheae. Microtubules were not observed after colchicine or vinblastin treatment. Hence, one requirement for tracheal migration in cultured wing disks is the presence of oriented microtubules.

Several laboratories have succeeded in inducing metamorphosis of cultured *Drosophila* disks with ecdysone. FRISTROM et al. (1969) reported that leg disks of *Drosophila* evaginated *in vitro* in response to α-ecdysone; and that this evagination was prevented by a juvenile hormone mimic. Complete metamorphosis of cultured *Drosophila* disks was obtained by MANDARON (1970). He demonstrated that not only will disks cultured *in vitro* with ecdysone evaginate, but they will also produce chitinized epidermal structures. It should be noted that MANDARON's preparations included the cerebral complex. KURODA (1969, 1970) found that *Drosophila* eye disks cultured in a chemically defined medium produced ommatidia in response to 20-

hydroxy-ecdysone and various ecdysone analogues of which rubrosterone was the most effective.

2. Competence of the Disks to Respond to Ecdysone

The ability of wing disks of various aged Lepidopteran larvae to respond to ecdysone has been examined in our laboratory (OBERLANDER, 1969a). We found that wing disks from larvae on each day of the last larval instar of *Galleria* initiated metamorphosis if cultured with α-ecdysone. However, the response of older disks was more consistent. Disks from both young and old larvae responded equally well to α-ecdysone if the medium was conditioned with fat body of any age (RICHMAN and OBERLANDER, 1971). Hence, the fat body may play an important role in the acquisition of competence by the disks. Mature disks in which the tracheae had already begun to migrate into the lacunae continued to develop in the absence of additional α-ecdysone. Similar observations were made by AGUI et al. (1969), who noted that wing disks from last instar *Chilo* larvae responded best to 20-hydroxyecdysone at a critical age, not at all in young disks, and became independent of additional hormone in mature disks.

3. Differences between Effects of α-ecdysone and 20-Hydroxyecdysone

There is still much to learn about the chemistry of the ecdysones. Nevertheless, two of the three ecdysones known to be indigenous to insects have been tested *in vitro*, and it is appropriate to ask whether these hormones produce identical or different effects. MANDARON (1970) reports complete morphogenesis of *Drosophila* disks *in vitro* in response to α-ecdysone, although it has been suggested that 20-hydroxyecdysone is the dominant ecdysone in the Diptera. FRISTROM et al. (1969) tested the effects of 20-hydroxyecdysone on cultured *Drosophila* disks, but obtained only evagination. AGUI et al. (1969) had excellent results with 20-hydroxyecdysone on cultured *Chilo* disks, but did not examine the effects of α-ecdysone. OBERLANDER (1969a) tested both α-ecdysone and 20-hydroxyecdysone, at comparable concentrations, on cultured *Galleria* wing disks. Both hormones induced tracheal migration, but 20-hydroxyecdysone usually caused a hypertrophy of the disks, while α-ecdysone induced normal elongation of the disks. It would be of interest to test various doses of the two ecdysones separately and in combination. A further difference between the effects of 20-hydroxyecdysone and α-ecdysone will be discussed in the section on proliferation.

4. α-ecdysone: Trigger or Sustained Stimulus?

It is important to determine whether α-ecdysone triggers metamorphosis *in vitro*, or whether it is required as a sustained stimulus. Wing disks from last instar *Galleria* larvae were cultured with α-ecdysone for varying times, and then placed in hormone-free medium (OBERLANDER, 1969a). A three day exposure to α-ecdysone was sufficient to induce tracheal migration in 100% of the cultured disks within six days after their removal into hormone-free medium. However, elongation occurred in 100% of the disks after six days of culture, but only in the continuous presence of α-ecdysone. Thus, these two phases in the development of the disks had different requirements for α-ecdysone. As mentioned earlier inhibition of tracheal migration did not prevent ecdysone-induced elongation. Hence, these processes are both dependent upon ecdysone, but are not dependent upon each other. From these observations I conclude that ecdysone is required as a sustained stimulus of metamorphosis.

Additional evidence that ecdysone is required as a sustained stimulus was provided by the observations of FALTUS and OBERLANDER (1970) that the development of pulsating regions in male *Drosophila* genital disks cultured in the abdomens of adult flies was enhanced by repeated doses of α-ecdysone. Similar observations were made by POSTLETHWAIT and SCHNEIDERMAN (1970) on the effects of 20-hydroxyecdysone on leg disks of *Drosophila* cultured *in vivo*. If ecdysone only triggered metamorphosis a dose which was adequate to begin the process should have been sufficient to cause continued independent development. This was not observed in these investigations thus far. However, it may well be that some phases of metamorphosis require only brief exposure to the hormone.

5. α-ecdysone Cofactors

A "macromolecular factor" has been reported in the haemolymph of *Cynthia* pupae to be necessary for spermatogenesis of pupal testes cultured *in vitro* with ecdysone (WILLIAMS and KAMBYSELLIS, 1969). This finding offered encouragement that such factors might improve imaginal disk development *in vitro* as well. RICHMAN and OBERLANDER (1971) cultured wing disks from the last larval instar of *Galleria in vitro* with and without larval fat body. α-ecdysone was added to the cultures after two days. The fat body increased the percentage of disks which responded to the hormone and reduced the time for this to occur. For example, wing disks taken from the fifth day of the last larval instar responded to 1 μg of α-ecdysone with tracheal migration in 30% of the disks eleven days after the addition of the hormone. None of these disks elongated. By contrast, a matched set of disks cultured under the same conditions, but with fat body, had tracheal migration in 90% of the disks after only five days, and elongation in 100% of the disks at nine days. If the medium was preconditioned with fat body which was removed prior to the addition of α-ecdysone the enhanced response did not occur. However, if fat body was cultured with the hormone for two days and removed before the addition of the disks to this medium, an improved response was obtained. Hence, the synergistic effect of fat body and α-ecdysone could be obtained only when the two were cultured simultaneously for at least two days. The "fat body factor" produced in response to α-ecdysone has recently been characterized as a dialyzable, pronase insensitive substance, which is stable to 50°C for 30 min and 100°C for 5 min. Its behavior in thin layer chromatography suggests that it may be a steroid (BENSON and OBERLANDER, in preparation). It appears that the fat body of *Galleria* plays a critical role in the response of wing disks to α-ecdysone. That the fat body improves the development of disks *in vitro* by altering the ecdysone molecule itself is a hypothesis that requires direct testing.

IV. Hormonal Control of Proliferation in Imaginal Disks

A. In vivo Experiments

As early as 1943 BODENSTEIN demonstrated that growth of imaginal disks could be influenced by hormones. He examined the growth of leg and eye disk pairs from *Drosophila melanogaster* and *Drosophila virilis* which were cultured in adult flies. One member of the pair was transplanted with 2—4 ring glands, while the matched disks were transplanted alone. BODENSTEIN observed that the disks transplanted to hosts with larval ring glands had grown larger than their matched disks in the control hosts. He also noted that disks transplanted into female adult hosts would grow some even in the absence of transplanted ring glands.

Subsequently, HADORN (1963) succeeded in long term culture of *Drosophila* imaginal disks *in vivo* by serial transplantations into adult hosts (see GEHRING, this volume). The disks grew best in fertilized females, less in virgin females and least in males (HADORN and GARCIA-BELLIDO, 1964). Whether the proliferation of these disks cultured *in vivo* is dependent upon hormones is not clear from these experiments, since we do not know whether fertilized females are better hosts because of an altered hormonal milieu or because of changes in other constituents of the haemolymph.

Experiments on the regeneration of imaginal disks in Lepidoptera suggest that there may be a feedback between the growing disks and the endocrine system. KROEGER (1958) and MUTH (1961) reported that molting in larvae of *Ephestia kühniella* was delayed when imaginal disks were implanted into the body cavity where they produced a second disk. POHLEY (1960, 1967) and MADHAVAN and SCHNEIDERMAN (1969) observed similar delays when larval disks were extirpated from larvae of either *Ephestia* or *Galleria*. MADHAVAN and SCHNEIDERMAN noted that the delay in molting was directly related to the number of disks which were removed. These authors performed careful control experiments in which the integument of the *Galleria* larvae was wounded, or a wing disk removed but the wound site cauterized to prevent regeneration. Pupation was not significantly delayed by either of these control procedures. Comparable observations in Diptera were made by URSPRUNG and HADORN (1962), who injected dissociated *Drosophila* imaginal disks into larvae. The cells proliferated extensively and led to a substantial delay of puparium formation.

The interactions between the endocrine system and the regenerating disks were considered by POHLEY (1961) who established that "the regeneration period ceases earlier in the presence of prothoracic glands than in their absence". He goes on to suggest that the "regenerating organs reduce the level of prothoracic gland hormone by using it up or by inactivation". Direct evidence that imaginal disk regeneration depends upon ecdysone was presented by MADHAVAN and SCHNEIDERMAN (1969). Only in 10% of the cases did the disks removed from larvae ligated behind the prothoracic glands become replaced by regenerated disks, compared to 75% in the unligated larvae. Injection of 10 μg/gm of α-ecdysone caused 65% of the ligated larvae to regenerate disks. The authors conclude from this that "ecdysone is necessary for the regeneration of imaginal disks in *Galleria*". The mechanism of delay in molting is left as an open question, and MADHAVAN and SCHNEIDERMAN suggest that the delay could be explained not only by an effect of the disks on ecdysone or the prothoracic glands, but also by an effect of the regenerative process on the responsiveness of other tissues to the molting hormone.

Several workers have used the incorporation of tritiated thymidine by imaginal disk cells as a parameter of cellular proliferation, since the disk cells do not become polyploid. BERREUR (1965) demonstrated that DNA synthesis in imaginal disks of *Calliphora* depended upon the presence of the lateral cells of the ring gland. Since ligature of *Calliphora* larvae to remove the ring glands would also trap some imaginal disks anterior to the ligature, BERREUR used a technique of POSSOMPÈS (1953) to remove the ring glands surgically. A radioautographic examination of tritiated thymidine incorporation into imaginal disk nuclei revealed that extirpation of the ring glands had inhibited DNA synthesis. The control level of incorporation of tritiated-thymidine was restored by implanting only that portion of the ring gland (lateral cells) which is responsible for molting hormone activity.

If the above experiments persuade us that growth of imaginal disks is controlled by ecdysone, how are we to explain the continuous growth of the disks during larval life when ecdysone is present cyclically (BODENSTEIN, 1957; EASSA, 1953; KRISHNAKUMARAN et al., 1967)? As a first step in answering the question we should know whether there are any differences between the cell cycles of young and mature disks.

An analysis of this sort was made on young and mature disks in *Ephestia* (LÖBBECKE, 1969). Using a double labeling technique with ^{3}H-thymidine injection followed by ^{14}C-thymidine he showed that wing disks during the first seven days of the last larval instar had an S-phase which lasted twelve hours, whereas in the prepupal period DNA was synthesized during a 5.8 h period of the cell cycle. LÖBBECKE suggested that this shift in the duration of the S-phase is controlled by ecdysone, but the hypothesis was not directly tested. Whether juvenile hormone plays a role in determining the nature of the cell cycle of young disks remains to be investigated.

B. In vitro Experiments

There have been few reports of hormone induced growth of imaginal disks *in vitro*. HORIKAWA (1960) noted that cultured eye disks of *Drosophila* grew in response to a crude extract (from Bombyx pupae) with molting hormone activity. Growth was also obtained when the eye disks were cultured with five or more cephalic complexes. OBERLANDER and FULCO (1967) reported that wing disks from the last larval instar of *Galleria* grew in response to α-ecdysone added to a chemically defined medium.

The effects of α-ecdysone and 20-hydroxyecdysone on the incorporation of ^{3}H-thymidine and ^{3}H-deoxycytidine in to DNA have been examined by OBERLANDER (1969a, b, 1971) using radioautographic techniques. The incorporation of ^{3}H-thymidine drops to zero in wing disks which have been cultured without hormone for 48 h. This drop occurs even in the presence of 3 μg/ml of 20-hydroxyecdysone. On the other hand wing disks continue to incorporate ^{3}H-thymidine when cultured in the presence of 3 μg/ml of α-ecdysone. Increasing the titer of α-ecdysone or the time of exposure to the hormone results in increased incorporation of the isotope. There is a considerable lag between addition of hormone and the turning on of DNA synthesis in disks which have been first cultured without hormone for 48 h. After 10 h exposure to the α-ecdysone the disks were removed to hormone free medium where they failed to initiate DNA synthesis even after an additional 66 h. 24 h exposure to α-ecdysone was sufficient to induce DNA synthesis, which then depended upon the continuous presence of the hormone. If the disks are removed to hormone free medium at any time after 24 h DNA synthesis stops within a few hours (OBERLANDER, unpublished observations) (see Fig. 1).

Further experiments revealed that concentrations of 20-hydroxyecdysone ranging from 0.03—30.0 μg/ml of medium failed to stimulate DNA synthesis in wing disks cultured with hormone for 48 h. Disks cultured in the continuous presence of 3 μg/ml of 20-hydroxyecdysone for six days did not incorporate ^{3}H-thymidine after the second day of culture. Control disks without hormone gave the same results. Despite this evidence it cannot be ruled out that a concentration and time yet to be tried would elicit a DNA synthesis response to 20-hydroxyecdysone. It was soon discovered, however, that 20-hydroxyecdysone was not neutral in connection with DNA synthesis. If disks were cultured with both α-ecdysone and 20-hydroxyecdysone

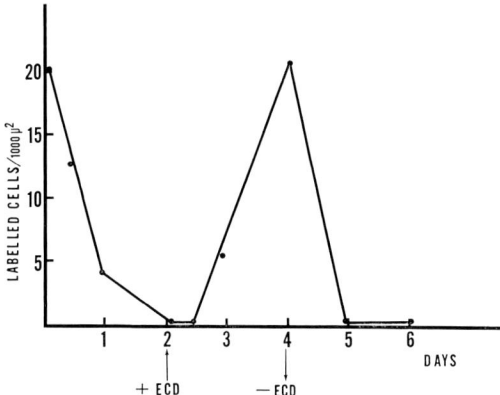

Fig. 1. Radioautographic evaluation of ^{3}H-thymidine incorporation into wing disks from the last larval instar of *Galleria mellonella* cultured *in vitro*. The data are reported as labelled cells (nuclei) per 1000 μ^2. The total number of cells in this area varies slightly because of the presence of the lacunae. Three μg/ml alpha-ecdysone was added on day two of culture. The disks were washed and placed in hormone free medium on day four (OBERLANDER, 1969b, and unpublished observations)

(3 μg/ml α-ecdysone and either 3 μg/ml or 30 μg/ml of 20-hydroxyecdysone) incorporation of ^{3}H-deoxycytidine was reduced to 10% of the expected level. α-ecdysone induced DNA synthesis was markedly reduced even if the 20-hydroxyecdysone was added well after the cells had initiated DNA synthesis (OBERLANDER, 1972).

To date there have been no clearcut demonstrations of an effect of natural juvenile hormone on proliferation in disks which are cultured *in vitro*. FRISTROM et al. (1969) observed that high doses (1 mg/ml) of a synthetic mixture with juvenile hormone activity reduced ^{3}H-thymidine incorporation in cultured *Drosophila* disks. There is an obvious need for work on the effects of pure juvenile hormone on the cell cycle in cultured imaginal disks.

V. Hormonal Control of RNA and Protein Synthesis in Imaginal Disks

A. In vivo Experiments

The interactions between hormones and RNA and protein synthesis have been the subject of numerous investigations by insect endocrinologists. However, there have been only a few papers devoted to the effects of ecdysone and juvenile hormone on RNA and protein synthesis in imaginal disks. Biochemical aspects of these studies are considered in detail by FRISTROM (this volume).

RNA synthesis has been examined by both radioautographic and biochemical techniques in the imaginal disks of larvae of *Samia cynthia ricini*. KRISHNAKUMARAN et al. (1965) and BERRY et al. (1967) used radioautography to follow the incorporation of ^{3}H-uridine into wing imaginal disks during the fourth larval instar. They observed that incorporation in the disks was always high relative to other tissues, but that there were some variations which were not obviously related to the molting cycle. The limitations inherent in quantitative work with radioautography did not permit more detailed analysis. PATEL and MADHAVAN (1969) used biochemical techniques to measure the incorporation of ^{3}H-uridine into wing disks during the third, fourth and fifth instars. The data were reported as disintegrations/min in perchloric acid precipitable material per µg of dry weight of disks. The incorporation of ^{14}C-protein hydrolysate into protein was followed in the same way. They found that increasing RNA and protein synthesis was correlated with rising ecdysone titer. The ecdysone titer was not measured in *Cynthia*, but was assumed to be the same as that in *Bombyx mori* (SHAAYA and KARLSON, 1965b). PATEL and MADHAVAN then injected α-ecdysone into larvae on the third day of the fourth instar and noted that both RNA and protein synthesis increased sharply. However, if α-ecdysone was injected on the fifth day protein synthesis increased as before, but RNA synthesis increased only slightly. The effects of dodecyl methyl ether (a juvenile hormone mimic) were less consistent. In response to 10 µg/gm of dodecyl methyl ether RNA synthesis in three day old disks was about ten times greater than the level found in injected five day old larvae. At 100 µg/gm of the hormone the rate of RNA synthesis at the two ages examined were about equal, while at 1000 µg/gm the rate of RNA synthesis in five day old disks was now twice that of three day old disks. Protein synthesis was consistently higher in five day disks than in three day disks at each of the concentrations of dodecyl methyl ether tested. Surprisingly, when both hormones were injected together there was little or no stimulation of RNA synthesis and a reduced effect on protein synthesis. PATEL and MADHAVAN concluded that an injected hormone had its greatest effect on RNA and protein synthesis when the endogenous level of that hormone was lowest. They state that "the imaginal disks are targets of both juvenile hormone and ecdysone", although *in vitro* experiments are required to determine whether RNA and protein synthesis in the disks are directly affected by these hormones. The antagonistic behaviour of α-ecdysone and the juvenile hormone mimic is puzzling, but has since been confirmed elsewhere (CONGOTE et al., 1969).

B. In vitro Experiments

The first publication on the effects of hormones on protein synthesis in larval imaginal disks cultured *in vitro* has yet to appear. BENSON and OBERLANDER (in preparation) examined the incorporation of ^{3}H-valine into TCA precipitable protein in wing disks of the last larval instar of *Galleria mellonella*. When disks were cultured in hormone free GRACE's medium (without haemolymph) incorporation of the ^{3}H-valine steadily dropped for three days and remained at a low level thereafter. The addition of α-ecdysone (3 µg/ml) to the cultures on the fourth day restored the initial level of protein synthesis by causing a three to five-fold increase in the level of incorporation of the isotope. WYATT and WYATT (1971) report that 20-hydroxyecdysone (5 µg/ml) elevates ^{3}H-leucine incorporation in *Cecropia* pupal wing tissue which is cultured *in vitro*. Hence, it appears likely that the ecdysones do directly stimulate protein synthesis in imaginal disks.

WYATT and WYATT (1971) observed an interesting difference between the effects of α-ecdysone and 20-hydroxyecdysone on RNA synthesis in cultured *Cecropia* pupal wing tissue. α-ecdysone produced a 30% increase on the level of incorporation of ^{14}C-uridine into RNA, although this was not observed in every case. The same level of 20-hydroxyecdysone (5 μg/ml) produced a 65% stimulation consistently. This stimulating effect of 20-hydroxyecdysone has been analyzed in some detail by FRISTROM et al. (1969) and RAIKOW and FRISTROM (1971). Imaginal disks of third instar *Drosophila melanogaster* were mass isolated and cultured in either ROBB's or SCHNEIDER's tissue culture medium. Concentrations of 20-hydroxyecdysone ranging from 0.01 μg/ml to 10.0 μg/ml stimulated the incorporation of ^{3}H-uridine into RNA. The uptake of ^{3}H-guanosine by the disks was increased by hormone, but this was insufficient to account for all of the increased incorporation of ^{3}H-guanosine into RNA. The authors conclude from this that possibly the "increase in incorporation of precursors into RNA results in part from an increase in RNA synthesis".

VI. Conclusions

Problems surrounding the growth and metamorphosis of imaginal disks have become a meeting ground for endocrinologists and developmental biologists. Despite early work in which BODENSTEIN, HADORN, VOGT and others (see Sections II and III) perceived the importance of imaginal disks as targets of insect hormones, modern insect endocrinology had all but forgotten the central position of the disks in the metamorphosis of many insects. This situation has been corrected recently as a result of three important developments: 1) The dramatic work of HADORN (1963) and his colleagues on the long-term culture of *Drosophila* disks *in vivo* has not only provided substantive contributions to developmental biology, but has also refocused the attention of many scientists on the imaginal disks. 2) The isolation, identification, synthesis and availability of pure insect hormones has made it possible for the endocrinologist to test known amounts of known hormones; and has thereby freed him from dependence upon transplantation of glands or crude extracts. 3) The successful culture of insect tissues *in vitro* (see review by MARKS, 1970) has provided a powerful tool for the analysis of interactions between hormones and various tissues under well defined conditions.

Experiments cited in this chapter on imaginal disk development *in vitro* demonstrate that the disks are direct targets of the ecdysones (Sections III and IV). It appears that α-ecdysone and 20-hydroxyecdysone have distinct effects on the disks (OBERLANDER, 1969a; 1971; WYATT and WYATT, 1971). In this connection it is interesting that MARKS (1971) has found that α-ecdysone was more effective than 20-hydroxyecdysone in initiating morphogenesis of cultured cockroach leg regenerates, and that the reverse was true for cuticle deposition. Whether these observations reflect physiological mechanisms operating *in vivo* is a matter for speculation.

It is to be regretted that the chemistry of insect hormones and the hormonal control of imaginal disk development have not been studied in the same species. Hence, investigations of the actions of hormones on imaginal disks can only assume that the titer (and identity) of hormones in the test insect is similar to that of *Bombyx mori*, *Manduca sexta* or *Calliphora*. Even in these species the chemistry of the ecdysones is imperfectly known.

Among the many problems associated with the hormonal control of imaginal disk development the following warrant particular attention. First of all, the biochemistry of hormone action on the imaginal disks requires thorough investigation. We know only that α-ecdysone can induce DNA synthesis in cultured disks, and that both α-ecdysone and 20-hydroxyecdysone can stimulate RNA and protein synthesis (Sections IV and V, and FRISTROM's chapter). Secondly, mechanisms of inhibition of metamorphosis of imaginal disks by pure juvenile hormone need to be studied. Finally, we do not know whether the proliferation and biosynthetic activity of premetamorphic disks are under hormonal control (EASSA, 1953; KRISHNAKUMARAN et al., 1967; PATEL and MADHAVAN, 1969). I believe that the use of *in vitro* culture methods will greatly facilitate the solution of these problems.

References

AGUI, N., YAGI, S., FUKAYA, M.: Effects of ecdysterone on the *in vitro* development of wing discs of rice stem borer, *Chilo supressalis*. Appl. Entomol. Zool. 4, 158—159 (1969).

BERKOFF, C. E.: The chemistry and biochemistry of insect hormones. Quart. Rev. 23, 372—391 (1969).

BERREUR, P.: Étude éxperimentale de l'action de l'hormone de mue sur l'évolution des acides nucleiques au cours de la métamorphose de *Calliphora erythrocephala* (Meig), insecte, diptère. Arch. Zool. Exp. 106, 531—624 (1965).

BERRY, S. J., KRISHNAKUMARAN, A., OBERLANDER, H., SCHNEIDERMAN, H. A.: Effects of hormones and injury on RNA synthesis in saturniid moths. J. Insect Physiol. 13, 1511—1537 (1967).

BODENSTEIN, D.: Hormones and tissue competence in the development of *Drosophila*. Biol. Bull., Woods Hole 84, 34—58 (1943).

BODENSTEIN, D.: Humoral dependence of growth and differentiation in insects. In: SCHEER, A., (Ed.): Recent Advances in Invertebrate Physiology, pp. 197—211. Eugene: Oregon University Press 1957.

BOWERS, W. S.: Juvenile hormones. In: JACOBSON, M., CROSBY, D. G. (Eds.): Naturally Occurring Insecticides, pp. 307—332. New York: Marcel Dekker, Inc. 1971.

BURDETTE, W. J., HANLY, E. W., GROSCH, H.: The effect of ecdysone on the maintenance and development of ocular imaginal discs *in vitro*. Texas Rep. Biol. Med. 26, 173—180 (1968).

BURTT, E. T.: On the corpora allata of dipterous insects II. Proc. roy. Soc. B 126, 210—223 (1938).

BUTENANDT, A., KARLSON, P.: Über die Isolierung eines Metamorphosehormons der Insekten in kristallisierter Form. Z. Naturforsch. 9b, 389—391 (1954).

CHADWICK, L. E.: Removal of prothoracic glands from the nymphal cockroach. J. exp. Zool. 131, 291—306 (1956).

CHERBAS, L., CHERBAS, P.: Distribution and metabolism of alpha-ecdysone in pupae of the silkworm *Antheraea polyphemus*. Biol. Bull. Woods Hole 138, 115—128 (1970).

CONGOTE, L. F., SEKERIS, C. E., KARLSON, P.: On the mechanism ot hormone action XIII stimulating effects of ecdysone, juvenile hormone, and ions on RNA Synthesis in fat body cell nuclei from *Calliphora erythrocephala* isolated by a filtration technique. Exp. Cell Res. 56, 338—346 (1969).

DAHM, K. H., TROST, B. M., RÖLLER, H.: The juvenile hormone V. Synthesis of the racemic juvenile hormone. J. Amer. chem. Soc. 89, 5292—5294 (1967).

EASSA, Y. E. E.: The development of imaginal buds in the head of *Pieris brassicae* Linn (lepidoptera) Trans. roy. Ent. Soc. (Lond.) 104, 39—50 (1953).

FALTUS, F., OBERLANDER, H.: Ecdysone induced differentiation of pulsating regions in genital imaginal disks after culture *in vivo*. Dros. Inf. Serv. 45, 155 (1970).

FEIR, D., WINKLER, G.: Ecdysone titres in the last larva and adult stages of the milkwede bug. J. Insect Physiol. **15**, 899—904 (1969).
FRISTROM, J. W., RAIKOW, R., PETRI, W., STEWART, D.: In vitro evagination and RNA synthesis in imaginal discs of *Drosophila melanogaster*. In: HANLY, E. W. (Ed.): Problems in Biology: RNA in Development, pp. 381—401. Salt Lake City: University of Utah Press 1969.
FUGIO, Y.: Studies of the development of eye-antennal discs of Drosophila melanogaster in tissue culture II. Effects of substances secreted from the cephalic complexes upon eye-antennal discs of mutant strains. Jap. J. Genet. **37**, 110—117 (1962).
FUKUDA, S.: Induction of pupation in silkworm by transplanting the prothoracic gland. Proc. Imp. Acad. Japan **16**, 414—416 (1940).
FURLENMEIER, A., FURST, A., LANGEMANN, A., WALDVOGEL, G., HOCKS, P., KERB, U., WIECHERT, R.: Zur Synthese des Ecdysons. Helv. chim. Acta **50**, 2387—2396 (1967).
GALBRAITH, M. N., HORN, D. H. S., THOMPSON, J. A., NEUFELD, G. J., HACKNEY, R. J.: Insect moulting hormones: crustecdysone in *Calliphora*. J. Insect Physiol. **15**, 1225—1233 (1969).
GERSCH, M., STURZEBECHER, J.: Weitere Untersuchungen zur Kennzeichnung des Activations-Hormons der Insektenhäutung. J. Insect Physiol. **14**, 87—96 (1968).
GILBERT, L. I., SCHNEIDERMAN, H. A.: The content of juvenile hormone and lipid in lepidoptera: sexual differences and developmental changes. Gen. comp. Endocrin **1**, 453—472 (1961).
GIRARDIE, A.: Action de la pars intercerebralis sur le developpement de *Locusta migratoria* L. J. Insect Physiol. **10**, 599—609 (1964).
GOTTSCHEWSKI, G.: Morphogenetische Untersuchungen an *in vitro* Wachsenden Augenanlagen von *Drosophila melanogaster*. Wilhelm Roux' Arch. Entwickl.-Mech. Org. **152**, 204—229 (1960).
HADORN, E.: An accelerating effect of normal "ring-glands" on puparium formation in lethal larvae of *Drosophila melanogaster*. Proc. nat. Acad. Sci. (Wash.) **23**, 478—484 (1937).
HADORN, E.: Differenzierungsleistungen wiederholt fragmentierter Teilstücke männlicher Genitalscheiben von *Drosophila melanogaster* nach Kultur *in vivo*. Develop. Biol. **7**, 617—629 (1963).
HADORN, E., GARCIA-BELLIDO, A.: Zur Proliferation von *Drosophila*-Zellkulturen im Adultmilieu. Rev. suisse de Zool. **71**, 576—582 (1964).
HADORN, E., NEEL, J. V.: Der hormonale Einfluß der Ringdrüse (Corpus allatum) auf die Pupariumbildung bei Fliegen. Wilhelm Roux' Arch. Entwickl.-Mech. Org. **138**, 281—304 (1938).
HOFFMEISTER, H.: Ecdysteron, ein neues Häutungshormon der Insekten. Angew. Chem. **78**, 269—270 (1966).
HORIKAWA, M.: Developmental genetic studies of tissue cultured eye-antennal discs of *Drosophila melanogaster* II. Effects of the metamorphic hormone (cephalic complex) upon growth and differentiation of eye-antennal discs and strain differences in relation to the metamorphic hormone. Jap. J. Genet. **35**, 76—83 (1960).
HORIKAWA, M., SUGAHARA, M.: Studies on the effects of radiation on living cells in tissue culture I. Radiosensitivity of various imaginal discs and organs in larvae of *Drosophila melanogaster*. Radiat. Res. **12**, 266—275 (1960).
HORN, D. H. S.: The Ecdysones. In: JACOBSON, M., CROSBY, D. G. (Eds.): Naturally Occurring Insecticides, pp. 333—462. New York: Marcel Dekker Inc. 1971.
ICHIKAWA, M., ISHIZAKI, H.: Protein nature of the brain hormone of insects. Nature (Lond.) Nature (Lond.) **198**, 308—309 (1963).
ICHIKAWA, M., NISHIITSUTSUJI-UWO, J.: Studies on insect metamorphosis VII. Effect of the brain hormone on the isolated abdomen of the eri-silkworm, *Philosamia cynthia ricini*. Mem. Coll. Sci. Univ. Kyoto **27**, 9—15 (1960).
ISHIZAKI, H., ICHIKAWA, M.: Purification of the brain hormone of the silkworm *Bombyx mori*. Biol. Bull. Woods Hole. **133**, 355—368 (1967).
JOHNSON, W. S., CAMPBELL, S. F., KRISHNAKUMARAN, A., MEYER, A. S.: Total synthesis of the Racemic form of the second juvenile hormone (methyl 12-homo juvenate) from the cecropia silkmoth. Proc. nat. Acad. Sci. (Wash.) **62**, 1005—1009 (1969).

Joly, P.: Endocrinologie des Insectes. Paris: Masson et Cie. 1968.
Kaplanis, J. N., Thompson, M. J., Yamamoto, R. T., Robbins, W. E., Louloudes, S. J.: Ecdysones from the pupa of the tobacco hornworm, *Manduca sexta* (johannson). Steroids **8**, 605—623 (1966).
Karlson, P.: Biochemical studies on insect hormones. Vitam. Horm. **14**, 227—266 (1956).
Karlson, P., Bode, C.: Die Inaktivierung des Ecdysons bei der Schmeißfliege *Calliphora erythrocepahla* Meigen. J. Insect Physiol. **15**, 111—118 (1969).
King, D. S., Siddall, J. B.: Conversion of alpha-ecdysone to beta-ecdysone by crustaceans and insects. Nature (Lond.) **221**, 955—956 (1969).
Kobayashi, M., Kirimura, J., Saito, M.: Crystallization of the "brain" hormone of an insect. Nature (Lond.) **195**, 515—516 (1962).
Kobayashi, M., Yamazaki, M.: The proteinic brain hormone in an insect, *Bombyx mori* L (lepidoptera: Bombycidae). Appl. Ent. Zool. **1**, 53—60 (1966).
Kopec, S.: Experiments on metamorphosis of insects. Bull. Int. Acad. Sci. Cracovie. 57—60 (1917).
Kopec, S.: Studies on the necessity of the brain for the inception of insect metamorphosis. Biol. Bull. Woods Hole **42**, 323—342 (1922).
Krishnakumaran, A., Berry, S. J., Oberlander, H., Schneiderman, H. A.: Nucleic acid synthesis during insect development II Control of DNA synthesis in the cecropia silkworm and other saturniid moths. J. Insect Physiol. **13**, 1—58 (1967).
Krishnakumaran, A., Oberlander, H., Schneiderman, H. A.: Rates of DNA and RNA synthesis in various tissues during a larval moult cycle of *Samia cynthia ricini* (lepidoptera). Nature (Lond.) **205**, 1131—1133 (1965).
Krishnakumaran, A., Schneiderman, H. A.: Prothoracotrophic activity of compounds that mimic juvenile hormone. J. Insect Physiol. **11**, 1517—1532 (1965).
Kroeger, H.: Über Doppelbildungen in die Leibeshöhle verpflanzter Flügelimaginalscheiben von *Ephestia kühniella* Z. Wilhelm Roux' Arch. Entwickl.-Mech. Org. **150**, 401—424 (1958).
Kuroda, Y.: The effect of ecdysone analogues on the differentiation of eye-antennal discs cultured in a chemically defined medium. Dros. Inf. Serv. **44**, 99—100 (1969).
Kuroda, Y.: Differentiation of ommatidium-forming cells of *Drosophila melanogaster* in organ culture. Exp. Cell Res. **59**, 429—439 (1970).
Kuroda, Y., Yamaguchi, K.: The effects of the cephalic complex upon the eye discs of *Drosophila melanogaster*. Jap. J. Genet. **31**, 98—103 (1956).
Löbbecke, E. A.: Autoradiographische Bestimmung der DNS Synthesedauer von Zellen der Flügelimaginalanlage von *Ephestia kühniella*. Wilhelm Roux Arch. Entwickl.-Mech. Org. **162**, 1—18 (1969).
Lyonet, P.: Traité anatomique de la chenille qui ronge le bois de saule. La Haye (1762).
Madhavan, K., Schneiderman, H. A.: Hormonal control of imaginal disc regeneration in *Galleria mellonella* (lepidoptera). Biol. Bull. Woods Hole **137**, 321—331 (1969).
Mandaron, P.: Développement *in vitro* des disques imaginaux de la *Drosophile*. Aspects morphologiques et histologiques. Devel. Biol. **22**, 298—320 (1970).
Marks, E. P.: The action of hormones in insect cell and organ cultures. Genl. Comp. Endocrin. **15**, 289—302 (1970).
Marks, E. P.: The induction of molting in insect organ cultures. III International Colloquium of invertebrate Tissue Culture. Smolenice, Czechoslovakia (1971).
Meyer, A. S., Hanzmann, E., Schneiderman, H. A., Gilbert, L. I., Boyette, M.: The isolation and identification of the two juvenile hormones from the cecropia silkmoth. Arch. Biochem. Biophys. **137**, 190—213 (1970).
Moriyama, H., Nakanishi, K., King, D. S., Okauchi, T., Siddall, J. B., Hafferl, W.: On the origin and metabolic fate of alpha-ecdysone in insects. Genl. Comp. Endrocinol. **15**, 80—87 (1970).
Muth, F. W.: Untersuchungen zur Wirkungsweise der Mutante „kfl" bei der Mehlmotte *Ephestia kühniella*. Wilhelm Roux' Arch. Entwickl.-Mech. Org. **153**, 370—418 (1961).
Nayar, K. K.: Effect of extirpation of neurosecretory cells on the metamorphosis of *Iphitia limbata* Stal. Current Sci. **25**, 192—193 (1956).

Nöthiger, R., Oberlander, H.: Differentiation of pulsating regions in genital imaginal discs after culture *in vivo*. J. exp. Zool. **164**, 61–68 (1967).

Novak, V. J. A.: Insect Hormones. London: Methuen and Co. 1966.

Oberlander, H.: Effects of ecdysone, ecdysterone, and inokosterone on the *in vitro* initiation of metamorphosis of wing discs of *Galleria mellonella*. J. Insect Physiol. **15**, 297–304 (1969a).

Oberlander, H.: Ecdysone and DNA synthesis in cultured wing disks of the wax moth, *Galleria mellonella*. J. Insect. Physiol. **15**, 1803–1806 (1969b).

Oberlander, H.: Alpha-ecdysone induced DNA synthesis in cultured wing disks of *Galleria mellonella*: Inhibition by 20-isoecdysone. J. Insect. Physiol. **18**, 223–228 (1972).

Oberlander, H., Fulco, L.: Growth and partial metamorphosis of imaginal discs of the greater wax moth. *Galleria mellonella in vitro*. Nature (Lond.) **216**, 1140–1141 (1967).

Ohtaki, T., Milkman, R. D., Williams, C. M.: Dynamics of ecdysone secretion and action in the fleshfly *Sarcophaga peregrina*. Biol. Bull. Woods Hole **135**, 322–334 (1968).

Piepho, H.: Zur Frage der Bildungsorgane des Häutingswirkstoffes bei Schmetterlingen. Naturwissenschaften **35**, 94 (1948).

Patel, N., Madhavan, K.: Effects of hormones on RNA and protein synthesis in the imaginal wing disks of the Ricini silkworm. J. Insect Physiol. **15**, 2141–2150 (1969).

Pochon-Masson, J.: Structure et fonctions des infrastructures cellulaires dénommées "microtubules". Ann. Biol. **7**, 361–390 (1967).

Pohley, H. J.: Experimentelle Untersuchungen über die Steuerung des Häutungsrhythmus bei der Mehlmotte *Ephestia kühniella* Zellen. Wilhelm Roux' Arch. Entwickl.-Mech. Org. **152**, 182–203 (1960).

Pohley, H. J.: Interaction between the endocrine system and the developing tissue in *Ephestia kühniella*. Wilhelm Roux' Arch. Entwickl.-Mech. Org. **153**, 443–458 (1961).

Pohley, H. J.: Regeneration experiments und kritische Perioden. Wilhelm Roux' Arch. Entwickl.-Mech. Org. **158**, 341–357 (1967).

Possompès, B.: Recherches expérimentales sur le déterminisme de la métamorphose de *Calliphora erythrocephala* Meig. Arch. Zool exp. gen. **89**, 203–364 (1953).

Postlethwait, J. H., Schneiderman, H. A.: Induction of metamorphosis by ecdysone analogues: *Drosophila* imaginal discs cultured in vivo. Biol. Bull. woods Hole **138**, 47–55 (1970).

Raikow, R., Fristrom, J. M.: Effects of beta-ecdysone on RNA metabolism of imaginal disks of *Drosophila melanogaster*. J. Insect Physiol. **17**, 1599–1614 (1971).

Richman, K., Oberlander, H.: Effects of fat body on alpha-ecdysone induced morphogenesis in cultured wing disks of the wax moth, *Galleria mellonella*. J. Insect Physiol. **17**, 269–276 (1971).

Robbins, W. E., Kaplanis, J. N., Svoboda, J. A., Thompson, M. J.: Steroid metabolism in insects In Annual Rev. Entomol. **6**, 53–72 (1971).

Röller, H., Bjerke, J. S.: Purification and isolation of juvenile hormone and its action in lepidopteran larvae. Life Sci. **4**, 1617–1624 (1965).

Röller, H., Dahm, K. H.: The identity of juvenile hormone produced by corpora allata *in vitro*. Naturwissenschaften **57**, 454–455 (1970).

Röller, H., Dahm, K. H., Sweely, C. C., Trost, B. M.: The structure of the juvenile hormone. Angew. Chem. **79**, 190 (1967).

Schläpfer, T.: Der Einfluß des adulten Wirtsmilieus auf die Entwicklung von larvalen Augenantennen-Imaginalscheiben von *Drosophila melanogaster*. Wilhelm Roux' Arch. Entwickl.-Mech. Org. **54**, 378–404 (1963).

Schneider, I.: Differentiation of larval *Drosophile* eye-antennal discs *in vitro*. J. exp. Zool. **156**, 91–104 (1964).

Sehnal, F.: Influence of the corpus allatum on the développment of internal organs in *Galleria mellonella* L. J. Insect Physiol. **14**, 73–85 (1968).

Sengel, P., Mandaron, P.: Aspects morphologiques du développement *in vitro* des disques imaginaux de la *Drosophile*. C. R. Acad. Sci. Ser. D **268**, 405–407 (1969).

Shaaya, E., Karlson, P.: Der Ecdysontiter während der Insektenentwicklung — II Die postembryonale Entwicklung der Schmeißfliege, *Calliphora erythrocephala*. J. Insect Physiol. **11**, 65–69 (1965a).

Shaaya, E., Karlson, P.: Der Ecdysontiter während der Insektenentwicklung — IV. Die Entwicklung der Lepidopteren *Bombyx mori* und *Cerura vinula* L. Dev. Biol. **11**, 424—432 (1965b).

Siddall, J.S., Cross, A.D., Fried, J.H.: Steroids. CCXCII. Synthetic studies on insect hormones II. The synthesis of ecdysone. J. Amer. chem. Soc. **80**, 862 (1966).

Srivastava, U.S., Gilbert, L.J.: Juvenile hormone: Effects on a higher Dipteran. Science **161**, 61—62 (1968).

Srivastava, U.S., Gilbert, L.I.: The influence of juvenile hormone on the metamorphosis of *Sarcophaga bullata*. J. Ins. Physiol. **15**, 177—189 (1969).

Thompson, M.J., Kaplanis, J.N., Robbins, W.E., Yamamoto, R.T.: 20,26-dihydroxyecdysone, a new steroid with moulting hormone activity from the tobacco hornworm. *Manduca sexta* (Johannson). Chem. Commun **1967**, 650

Ursprung, H., Hadorn, E.: Weitere Untersuchungen über Musterbildung in Kombinaten aus teilweise dissoziierten Flügelimaginalscheiben von Drosophila melanogaster. Dev. Biol. **4**, 40—66 (1962).

van der Kloot, W.G.: Insect metamorphosis and its endocrine control. Amer. Zoologist **1**, 3—9 (1961).

Vogt, M.: Induktion von Metamorphoseprozessen durch implantierte Ringdrüse bei *Drosophila*. Wilhelm Roux' Arch. Entwickl.-Mech. Org. **142**, 129—182 (1942a).

Vogt, M.: Die „Puparisierung" als Ringdrüsenwirkung. Biol. Zbl. **62**, 149—154 (1942b).

Vogt, M.: Zur Kenntnis des larvalen und pupalen Corpus Allatum von *Calliphora*. Biol. Zbl. **63**, 56—71 (1943a).

Vogt, M.: Aus Produktion und Bedeutung metamorphosefördernder Hormone während der Larvenentwicklung von *Drosophila*. Biol. Zbl. **63**, 395—446 (1943b).

Vogt, M.: Inhibitory effects of the corpora cardiaca and of the corpus allatum in *Drosophila*. Nature (Lond.) **157**, 512 (1946).

Weismann, A.: Die nachembryonale Entwicklung der Musciden nach Beobachtungen an *Musca vomitoria* und *Sarcophaga carnaria*. Z. wiss. Zool. **14**, 187—336 (1864).

White, J.G.: Effects of colchicine and *Vinca* alkaloids on human platelets I. Influence on platelet microtubules and contractile function. Amer. J. Path. **53**,(2) 281—291 (1968).

Wigglesworth, V.B.: Function of the corpus allatum in insects. Nature (Lond.) **136**, 338 (1935).

Wigglesworth, V.B.: The function of the corpus allatum in the growth and reproduction of *Rhodenius prolixus*. Quart. J. micr. Sci. **79**, 91—119 (1936).

Wigglesworth, V.B.: Häutung bei Imagines von Wanzen. Naturwissenschaften **27**, 301 (1939).

Wigglesworth, V.B.: The determination of characters at metamorphosis in *Rhodnius prolixus* (Hemiptera). J. exp. Biol. **17**, 201—222 (1940).

Wigglesworth, V.B.: The Physiology of insect metamorphosis. London: Cambridge Univ. Press 1954.

Wigglesworth, V.B.: The hormonal regulation of growth and reproduction in insects, in Advances in Insect Physiology. pp. 248—336. New York: Academic Press 1964.

Williams, C.M.: Physiology of insect diapause II. Interaction between the pupal brain and prothroacic glands in the metamorphosis of the giant silkworm, *Platysamia cecropia*. Biol. Bull. Woods Hole **93**, 89—98 (1947).

Williams, C.M.: The juvenile hormone of insects. Nature (Lond.) **178**, 212 (1956).

Williams, C.M.: The juvenile hormone II. Its role in the endocrine control of molting, pupation and adult development in the cecropia silkworm. Biol. Bull. **121**, 572—585 (1961).

Williams, C.M.: Nervous and Hormonal Communication in insect development. Dev. Biol. Suppl. **3**, 133—150 (1969).

Williams, C.M., Law, J.H.: The juvenile hormone IV. Its extraction, assay and purification. J. Insect Physiol. **11**, 569—580 (1965).

Williams, C.M., Kambysellis, M.: *In vitro* action of ecdysone. Proc. nat. Acad. Sci. (Wash.) **63**, 231 (1969).

Wyatt, S.S., Wyatt, G.R.: Stimulation of RNA and protein synthesis in silkmoth pupal wing tissue by ecdysone *in vitro*. Gen. comp. Endocrin. **16**, 369—374 (1971).